Introducing Natural Resources

Other Titles in this Series:

Introducing Astronomy
Introducing Geology – A Guide to the World of Rocks
(Second Edition)
Introducing Geomorphology
Introducing Meteorology ~ A Guide to the Weather
Introducing Mineralogy
Introducing Oceanography
Introducing Palaeontology ~ A Guide to Ancient Life
Introducing Sedimentology
Introducing Stratigraphy (forthcoming)
Introducing Tectonics, Rock Structures and Mountain Belts
Introducing the Planets and their Moons
Introducing Volcanology ~ A Guide to Hot Rocks

For further details of these and other Dunedin Earth and Environmental Sciences titles see
www.dunedinacademicpress.co.uk

INTRODUCING
NATURAL RESOURCES

Graham Park

EDINBURGH ◆ LONDON

Published by
Dunedin Academic Press Ltd
Hudson House
8 Albany Street
Edinburgh EH1 3QB

London office:
352 Cromwell Tower
Barbican
London EC2Y 8NB

www.dunedinacademicpress.co.uk

ISBNs
9781780460482 (Paperback)
9781780465494 (ePub)
9781780465500 (Kindle)

British Library Cataloguing in Publication data
A catalogue record for this book is available from the British Library

Typeset by Makar Publishing Production, Edinburgh, Scotland
Printed in Poland by Hussar Books

Contents

List of illustrations and tables

Sourced illustrations

The following illustrations are reproduced by permission:
Shutterstock: figures 2.1A; 2.4A, B, C; 5.1 B; 5.2A, B, D; 5.3A, B; 5.4; 6.1; 7.1; 7.2A, B; 7.3; 8.1A, B, C, D; 8,2; 11.1A; 11.2A; 11.4.
Figure 5.1A: Science Photo Library.

The following illustrations have been adapted from published sources:
Figure 2.3A: Wikipedia/commons/e/e6/SolarSystemAbundances.
Figure 2.3B: Ahrens, L.H. (1969). The composition of stone meteorites: *Earth and Planetary Science Letters* 5, 3.
Figure 4.2A, C, D: Stanton, R.L. (1989). Ore Petrology, McGraw-Hill.
Figure 9.2: WikimediaCommons: Nat Gas Processing.
Figure 9.3: Delphi 234, via Wikimedia/commons.
Figure 10.3: Encyclopedia Britannica.inc, 2011.
Figure 10.4: Veizer, J. (1999). $^{87}Sr/^{86}Sr$, $\delta^{13}C$ and $\delta^{18}O$ evolution of Phanerozoic seawater. *Chemical Geology* 161, 59–88.
Figure 11.5: Renewable Energy Index, Jan. 2010.
Figure 12.2: Petit, J.R., Jouzel, J., Raynaud, D., Barkov, N.I., Barnola, J.-M., Basile, I., Bender, M., Chappellaz, J., Davis, M., Delaygue, G., Delmotte, M., Kotlyakov, V.M., Legrand, M., Lipenkov, V.Y., Lorius, C., Pepin, L., Ritz, C., Saltzman, E., and Stievenard, M. (1999). Climate and atmospheric history of the past 420,000 years from the Vostok ice core, Antarctica. *Nature* **399**: 429–436.
Figure 12.3: Exxon sea-level curve, e.g. Haq et al. (1987). *Science* 235, 156–1167.

Sources of other images:
Figure 12.1, table 12.1: US Geological Survey.

Preface

There is already an extensive literature both on natural resources and on the various associated topics such as climate change, reduction of biodiversity, destruction of habitat and so on, and it is not my intention to duplicate that. What I have attempted to do in this short book is to introduce the more important topics in a relatively simple manner, avoiding complex scientific terms and arguments, with the aim of informing the general reader about some of the most important issues of the day. Some previous knowledge of geology would be helpful in understanding parts of this book, but is not essential as terms that might be unfamiliar are highlighted in bold and defined in the Glossary.*

Over the many millennia that the human race has inhabited our planet, a use has been found for almost everything that is to be found on it. However, since the Industrial Revolution, many of the resources that we have come to rely on are being depleted, some at an alarming rate. Misuse of others, such as fossil fuels, is causing such damage to the environment that measures are being taken at an international level to restrict their use.

Earth is unique among the planets of the Solar System in providing not only an atmosphere and temperature range capable of sustaining life as we know it, but also accessible supplies of the various mineral resources that are now regarded as essential for us.

There is an important distinction between *renewable* and *non-renewable* types of natural resources. Non-renewable resources may be divided into four categories: 1) *metallic minerals*, including the ores of the well-known metals, but also those of uncommon and strategically important ones; 2) *non-metallic minerals* such as nitrates and phosphates; 3) economically important *rocks* such as limestone, salt and gypsum; and 4) *fossil fuels* (coal, oil and natural gas), which are obviously of great economic significance. The possibility of the exhaustion of certain non-renewable resources poses significant problems.

Renewable resources include all plant and animal life, together with some physical resources that are effectively infinite, such as solar energy, air and water. In practice, renewability may be limited and exhaustion of the resource foreseeable: for example plant and animal life is a renewable resource only if properly managed, hence the importance of maintaining biodiversity. Renewable energy resources include biomass, solar, geothermal, wind, hydro, tidal and wave power. These have both advantages and limitations as energy sources.

The environment in which all living things exist, termed the *biosphere*, includes both the atmosphere and the oceans. It is important to understand the relationship between these and the necessity of keeping them in balance. Human intervention has created dangerous disturbances to the environment that have caused great debate among the scientific community; the topic of global warming in particular, and its effects on climate, is now a matter of widespread public concern.

*Note: all terms initially highlighted in **bold** are defined in the Glossary at the end of the book.

Acknowledgements

I am indebted to Professor John Winchester for many helpful comments and suggestions that have resulted in significant improvements to this book; to an anonymous reviewer who was responsible for some important changes to Chapter 2, and to my wife, Sylvia for her unfailing support and for checking the manuscript for general readability

1 Introduction

What is a resource?

The Oxford Dictionary definition of a resource is: '*an available stock or supply that can be drawn on; an asset*; or '*a country's collective wealth*'. **Natural resources** are those that occur naturally, rather than those that have accumulated due to human activity. In considering the origin of the natural resources of our planet, it is necessary first to understand how the Solar System came into being, how the composition of the Earth itself was determined, and how its early history shaped the distribution of its natural resources and the evolution of its life forms. These topics are summarized in chapter 2.

Earth's natural resources fall into two basic categories: *renewable* and *non-renewable*. Non-renewable resources (chapters 3–9) are those of which the Earth has a finite supply, and which can become exhausted, whereas renewable resources could, in theory, be used indefinitely.

Resources and reserves

Economic geologists recognize a distinction between the **resources** of a useful or potentially useful substance such as a metallic ore mineral, and the **reserves** of that resource, which are those that have been calculated to exist and which are economically feasible to extract. Estimates of reserves frequently change due to the effects of continued extraction, new discoveries and price changes.

Non-renewable resources

The reason we are able to access useful natural resources of both minerals and rocks is that they have become concentrated by geological processes of one kind or another. Many elements are extremely rare and are only found in minute quantities in the commoner igneous rocks. Chapter 3 examines the various processes by which the concentrations of these elements are achieved.

The supply of physical resources, such as minerals, fossil fuels, etc., is obviously of great economic significance; the question of their possible exhaustion is in many cases all too obvious, and poses significant and pressing problems.

Non-renewable physical resources may be divided into four categories:
1. *Metallic mineral resources*: e.g., ores of iron, nickel, etc. (chapters 4–6).
2. *Non-metallic mineral resources*: e.g., nitrates, phosphates (chapter 7).
3. *Certain economically important rocks*: e.g., limestone, gypsum (chapter 8).
4. *Fossil fuels*: coal, oil and natural gas (chapter 9).

The factors that have to be taken into account in the case of a non-renewable resource are its genesis – the geological processes leading to its formation – its mode of occurrence; its abundance; the methods and ease of its extraction; the likelihood of its potential exhaustion; and any obvious 'social' problems such as toxicity and environmental pollution.

Non-renewable energy resources (chapter 9) are particularly important and a vigorous debate is taking place as to whether, and how much, we should limit our reliance on them.

Renewable resources

All plant and animal life, together with some physical resources that are effectively infinite, such as air, water, and solar energy, are, in theory, **renewable resources**. However 'renewable' in this context is relative rather than absolute, since renewability may, in practice, be limited, and exhaustion of the resource may be foreseeable due to human activity, habitat loss, or, over a longer time span, to evolutionary changes. An example of the possible limitations of a renewable resource is plant life. Because of over-exploitation or poor husbandry, certain plant species have become extinct or severely limited in their extent, adversely affecting potential uses in the future. As a result of this, although in theory the supply of a certain plant species may be infinitely renewable, in practice, due to poor management, it might become exhausted. Soil is another example of a resource that might at first sight seem to be infinitely renewable but, as can be seen especially in many developing countries, soil erosion and over-exploitation have severely depleted this resource.

Chapter 10 discusses the nature and composition of the atmosphere and the oceans, and their relationship with living organisms. The delicate balance between them, reflected in the current debate about excessive carbon dioxide release, is of immediate concern. Other current problems that are causing anxiety among biologists are the destruction of ecological habitats and the related decrease in biodiversity. The reduction in numbers and diversity of species, many of which are interrelated in ways that may not yet be understood, adversely affects the possibilities of their future exploitation as a resource.

Renewable energy resources such as *solar, geothermal, wind, wave* and *tidal energy* are particularly important and topical, and are discussed in chapter 11 in terms of their potential as energy sources and the limitations of their possible exploitation. These topics have assumed great urgency in recent years because of concerns over global warming.

Protecting the Planet

The final chapter concludes with a résumé of some of the more important problems facing us today. These include: over-exploitation of scarce resources, pollution of the environment by dangerous or toxic substances, degradation of the natural ecology, and perhaps the most pressing problem: the excessive release of greenhouse gases with their impact on global warming and climate change.

2   Origin and early history of the Earth

Origin of the Solar System

Theories about the origin of the Solar System have varied over the years; however, according to currently popular ideas, the Earth emerged 4.6 Ga[1] ago from a hot disc of gas and dust that formed a kind of nebular body with the Sun at its centre, extending out to the limits of the present Solar System. The minute particles making up this 'solar nebula', in addition to the lighter elements such as hydrogen and helium that form the overwhelming majority, also include a small proportion of the heavier elements, which are widely dispersed throughout the system. Those elements that are heavier than iron are thought to have been produced in the core of an ancient star that has subsequently exploded forming a **supernova**, sending clouds of heavier elements into space to be captured by bodies such as the Sun and the planets.

The particles within this solar nebula were subjected to a barrage of random collisions between neighbouring particles, resulting in the formation of clusters which, because of their gravitational attraction, gradually coalesced with others into kilometre-sized bodies known as **planetesimals**. In time, some of these grew by further accretion to become planets. Many of the remainder still exist and form the **asteroids**.

1 [Ga = Giga anni = 1000 million years]

The Sun

At the centre of the Solar System is the Sun, which contains most of the mass of the System, and consists of around 98% hydrogen and helium but also, crucially, small amounts of the denser elements, including the metals, which are now present in the outer regions of the Sun in roughly the same proportions as they are in the Earth. The energy of the Sun is produced by the **nuclear fusion** of hydrogen to form helium in the same process used in the hydrogen bomb.

The planets

There is enormous variation in the planets, depending mainly on their distance from the Sun (Fig. 2.1). The inner planets – Mars, Earth, Venus and Mercury – also known as the 'terrestrial' planets, are dense and 'rocky', containing a high proportion of **silicate** minerals and metals, whereas the much larger outer planets are mostly composed of gas – mainly hydrogen and helium in the case of Jupiter and Saturn, or of ice (e.g. frozen water, methane and ammonia) in the case of the outermost planets, Uranus and Neptune. Variation within the inner planets is largely due to temperature differences caused by their distance from the Sun. The innermost planet, Mercury, is too hot to be able to sustain an atmosphere, any initial surface liquids or gases having been lost long ago, whereas the outermost terrestrial planet, Mars, is too cold, and

any water is in the form of ice. Only Venus and Earth have dense atmospheres, although in the case of Venus, the surface temperatures are much

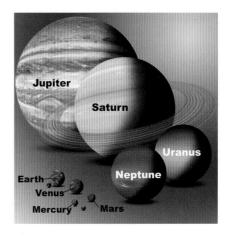

A

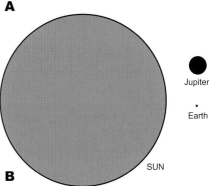

B

Figure 2.1 The Solar System
A Relative sizes of the planets; note the much larger outer planets: Jupiter, Saturn, Uranus and Neptune. Image credit: © Shutterstock, alexaldo. **B** The size of the Sun relative to that of the planets Jupiter and Earth.

higher (c.400°) than Earth's. The Earth is the only solar planet known to experience continuous geological activity in the form of plate tectonics and vulcanicity, and to possess surface water.

The Solar System also contains many smaller objects such as comets and asteroids – the latter concentrated in the asteroid belt, between the inner and outer planets, and in an outer belt beyond Neptune containing 'dwarf' planets such as Pluto, composed mainly of ice.

Formation of the elements

The elements found on Earth are the same as those observed in other parts of the Solar System and considered to be common to the whole Universe (Fig. 2.2). They are formed by the process of nucleosynthesis, which creates new,

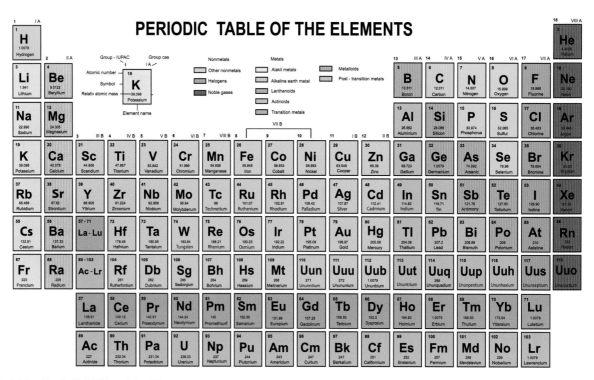

Figure 2.2 The Periodic Table of the Elements
The elements listed in order of increasing atomic weight: 1 hydrogen, H; 2 helium, He; 3 lithium, Li; 4 beryllium, Be; 5 boron, B; 6 carbon, C; 7 nitrogen, N; 8 oxygen, O; 9 fluorine, F; 10 neon, Ne; 11 sodium, Na; 12 magnesium, Mg; 13 aluminium, Al; 14 silicon, Si; 15 phosphorus, P; 16 sulphur, S; 17 chlorine, Cl; 18 argon, Ar; 19 potassium, K; 20 calcium, Ca; 21 scandium, Sc; 22 titanium, Ti; 23 vanadium, V; 24 chromium, Cr; 25 manganese, Mn; 26 iron, Fe; 27 cobalt, Co; 28 nickel, Ni; 29 copper, Cu; 30 zinc, Zn; 31 gallium, Ga; 32 germanium, Ge; 33 arsenic, As; 34 selenium, Se; 35 bromine, Br; 36 krypton, Kr; 37 rubidium, Rb; 38 strontium, Sr; 39 yttrium, Y; 40 zirconium, Zr; 41 niobium, Nb; 42 molybdenum, Mo; 43 technetium, Tc; 44 ruthenium, Ru; 45 rhodium, Rh; 46 palladium, Pa; 47 silver, Ag; 48 cadmium, Cd; 49 indium, In; 50 tin, Sn; 51 antimony, Sb; 52 tellurium, Te; 53 iodine, I; 54 xenon, Xe; 55 caesium, Cs; 56 barium, Ba; 57 lanthanum, La; 58 cerium, Ce; 59 praseodymium, Pr; 60 neodymium, Nd; 61 promethium, Pm; 62 samarium, Sm; 63 europium, Eu; 64 gadolinium, Gd; 65 terbium, Tb; 66 dysprosium, Dy; 67 holmium, Ho; 68 erbium, Er; 69 thulium, Tm; 70 ytterbium, Yb; 71 lutetium, Lu; 72 hafnium, Hf; 73 tantalum, Ta; 74 tungsten, W; 75 rhenium, Re; 76 osmium, Os; 77 iridium, Ir; 78 platinum, Pt; 79 gold, Au; 80 mercury, Hg; 81 thallium, Tl; 82 lead, Pb; 83 bismuth, Bi; 84 polonium, Po; 85 astatine, At; 86 radon, Rn; 87 francium, Fr; 88 radium, Ra; 89 actinium, Ac; 90 thorium, Th; 91 protactinium, Pa; 92 uranium, U; 93 neptunium, Np; 94 plutonium, Pu; 95 americium, Am; 96 curium, Cm; 97 berkelium, Bk; 98 californium, Cf; 99 einsteinium, Es; 100 fermium, Fm; 101–118 trans-fermium elements: only artificially produced. Image credit: © Shutterstock.

more complex, atomic nuclei from existing atomic or sub-atomic particles by means of various nuclear reactions. The first stage in this process is the creation of helium from hydrogen by nuclear fusion, which powers the energy of the Sun. Further stages in the creation of the heavier elements are thought only to occur in much larger stars. The heavier elements, above iron in the Periodic Table (Fig. 2.2), are attributed to the collapse of a supernova, as outlined above.

Figure 2.3A is an estimate of the cosmic abundance of all the elements in the Periodic Table as far as uranium (atomic number 92), the abundance being measured on a logarithmic scale relative to silicon, whose abundance is arbitrarily taken as 1,000,000. Hydrogen is 10,000 times more abundant, whereas uranium is 100 million times less abundant. With the exception of a few lighter elements (lithium, beryllium, boron and fluorine) the abundance gradually reduces through the table as the elements become heavier. This reflects the increasing difficulty of making them. The spikiness of the graph reflects the fact that elements

Figure 2.3 The abundance of the elements
A Cosmic abundance values relative to silicon = 10^6; only some of the commoner elements are named: *see* Fig. 2.2 for full list; each even-numbered element is slightly more abundant than the next odd-numbered one due to differences in atomic structure; this produces the jagged pattern on the graph. **B** Element abundances in the average carbonaceous chondrite meteorite compared with that of the Sun's atmosphere. Note that lithium is strongly enriched, while nitrogen and carbon are depleted. Some elements are not shown. In both A and B, the scales are logarithmic, so that each step is ×10 from the last.

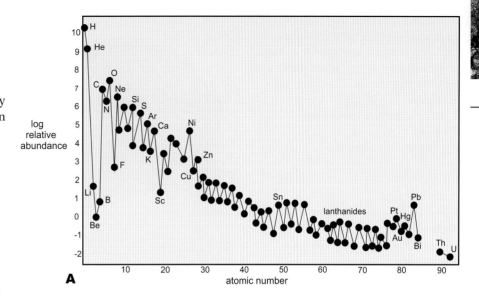

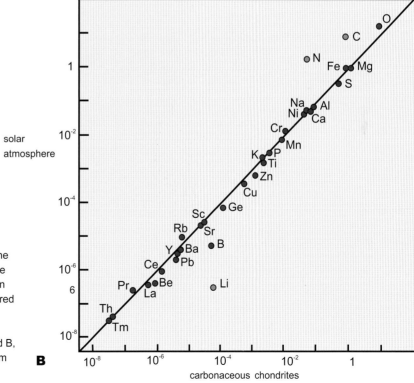

with odd atomic numbers are more difficult to synthesize, and are therefore slightly less abundant, than their even-numbered neighbours: thus nitrogen (7) is slightly less abundant than oxygen (8).

Further changes take place within the Earth due to radioactive decay of the original radioactive elements such as uranium and thorium, which eventually decay to lead, via several intermediate stages.

Meteorites

Meteorites are pieces of solid rock that have arrived at the surface of the Earth from outer space, and are an essential source of information about the origin and composition of the Earth. They represent a sample of the kind of material from which the Earth itself was formed. Almost all are about 4.6Ga old – presumed to be the same age as the Earth. Meteorites vary enormously in size; most range from small pebbles to large boulders (Fig. 2.4). The largest form huge craters and disintegrate on impact. Over 38,000 have been recorded. There are three main types: stony meteorites (including chondrites), iron meteorites and stony-iron meteorites.

Chondrites

Around 86% of meteorite finds are the type known as chondrites. These are important because they have not been modified by melting from their parent body and can therefore be assumed to represent the primitive material of the Solar Nebula. Over 90% of chondrites contain small spherical globules of frozen silicate melt, termed chondrules, contained within a crystalline matrix of silicate minerals

Figure 2.4 Meteorites
A Iron meteorite. Image credit: © Shutterstock, Chinellato Photo. **B** Cut and polished section showing iron-nickel intergrowth pattern in an iron meteorite from Sweden. Image credit: © Shutterstock, Albert Russ. **C** The famous Barringer meteorite impact crater, Arizona; the crater is 1km across and 200m deep. Image credit: © Shutterstock, ActionSportsPhotography.

plus variable amounts of metallic iron and nickel. The silicate minerals are those familiar from earthly igneous rocks: pyroxene, olivine and feldspar (see Appendix, Fig. A2). About 5% of chondrites sampled contain some carbon-based compounds, including amino-acids, which have given rise to speculation that such bodies may have transferred the essential ingredients for early life to Earth. A small number contain hydrous minerals attributed to their having acquired some ice particles. Although some specimens have experienced a degree of thermal metamorphism, none has been completely melted; however, some display evidence of localized melting assumed to be due to collisions, either with the Earth itself or with other bodies.

The relative abundance of the elements contained within chondrites corresponds to that calculated from the atmosphere of the Sun and other stars of our galaxy, as shown in Figure 2.3A and B, with the exception of the volatile elements: hydrogen, carbon, nitrogen and the 'inert' gases (helium, neon, argon, krypton, xenon and radon). This helps to explain the concentrations of the various mineral resources in the Earth, as will be seen.

Iron meteorites

The second meteorite type, the iron meteorites, making up only about 6% of finds, are composed of metallic alloys of iron and nickel. These are considered to belong to the cores of asteroids that were once molten, enabling the denser metallic material to sink towards the centre of these bodies; this is part of the process known as **planetary differentiation**, in which the constituents of a planetary body are separated out because of their different physical or chemical characteristics.

The third type of meteorite is the stony-iron type; this represents only about 1% of finds and includes two subtypes: **pallasites** and **mesosiderites**. Pallasites are thought to have originated in the boundary region between the stony and iron zones of the parent asteroid.

Mesosiderites are metal-bearing stony **breccias** of probable impact origin.

A very few meteorite finds are enriched in the lighter silicates and are similar to some of Earth's crustal rocks, such as **basalt**. They are thought to represent material from the outer 'crust' of an asteroid.

The Earth

The Earth's interior

The interior of the Earth is divided into three regions (Fig. 2.5A). There is an outer thin shell called the **crust**, which varies in thickness from about 7 kilometres beneath the oceans, to an average of about 35 kilometres in continental areas, reaching around 80 kilometres beneath certain young mountain belts. The innermost, approximately spherical, region, the core, extends from a depth of 2900 kilometres to the centre of the Earth, at a depth of 6500 kilometres. The region lying between the crust and the core is termed the **mantle**, and it is here that the processes that largely control what happens at the Earth's surface originate.

The composition of the crust is known in some detail, since material derived from all depths of the crust is directly accessible at the surface somewhere. There is a great difference between oceanic and continental crust. Oceanic crust is composed almost entirely of the volcanic rock basalt (and its coarser equivalents), whereas continental rocks are extremely varied in composition, including the whole range of igneous rock types (*see* Appendix, table A2) together with

all the different types of sedimentary and metamorphic rock derived from them. However, the average composition has been estimated to be broadly similar to that of granite.

The core is believed to consist mostly of metallic iron, with some other elements in addition, particularly nickel, corresponding to the composition of the iron meteorites. In the outer core, the metal is in a molten state but the inner core is solid. These dense constituents are thought to have become molten at an early stage in Earth's history and drained down towards the centre forming the core – the same process of planetary differentiation thought to explain the meteorite compositions.

The chemical composition of the mantle, taken as a whole, is considered to approximate to that of the average chondrite (see Fig. 2.3B). This composition, referred to as 'primitive mantle', is considered to be the starting point of all basaltic magmas,

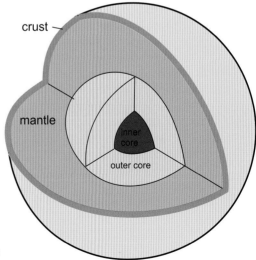

Figure 2.5A The interior of the Earth

whose compositions are often given, for convenience, in terms of their variation from either standard chondrite or primitive mantle models. There is good evidence that much of the upper mantle, above about 400km, is made of the ultrabasic rock **peridotite**. This material is known to melt to form basalt magma, and pieces of peridotite are found within some basalts. Moreover, in places where oceanic crust has been thrust onto continental margins, mantle material consisting of peridotite is found directly beneath the crustal rocks. Peridotite is composed mainly of the minerals olivine and pyroxene, which are silicates of iron and magnesium. Other minerals probably also present in mantle peridotite include feldspar, **garnet** and **spinels** (metallic oxides). All the other elements found in the crust, including the heat-producing radioactive elements, must also be present in small quantities, either within the main silicate minerals or in other compounds such as oxides, since the crust has been formed over time from magmas derived from the mantle. Below 400km, at the higher pressure, olivine is unstable and would be replaced by higher-pressure minerals such as spinel and garnet. Below 650km, all the upper-mantle silicate minerals are unstable and the lower mantle is thought to be composed of high-density minerals similar to the titanium oxide mineral, **perovskite**.

The atmosphere

The outer regions of the Earth, above the solid surface, consist of the **hydrosphere**, comprising the oceans, and the **atmosphere** (Fig. 2.5B). The lowest part of the atmosphere is termed the **troposphere** and consists of a mixture of

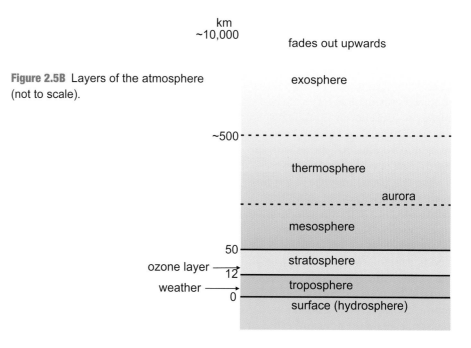

Figure 2.5B Layers of the atmosphere (not to scale).

gases, principally nitrogen and oxygen, together with minor (but important) amounts of carbon dioxide, methane and also the 'inert' gases, helium, neon, and argon. The concentration of these gases increases downwards. The troposphere, only about 12km thick, is the zone where the weather is formed.

Above the troposphere is the **stratosphere**, which includes the layer where some of the oxygen has been altered to **ozone** (O_3) by the Sun's radiation, and which protects Earth from the harmful effects of ultraviolet radiation. The stratosphere is around 40km thick and fades gradually upwards into the **mesosphere**, which in turn is replaced upwards by the **thermosphere**; this extends to between 500 and 1000km from the surface. The lower part of the thermosphere contains the **ionosphere**, where the phenomena of the **aurora borealis** and the **aurora australis** occur.

The outermost layer, the **exosphere**, extends up to about 10,000km, and fades into outer space. This layer contains extremely low densities of gas, mainly hydrogen and helium, whose molecules are so far apart that they may travel many kilometres before colliding with a neighbour!

Earth's age and source of heat

In the late nineteenth century, it was generally believed that the Earth was only 20–80 million years old, based on a calculation by Lord Kelvin, who assumed that Earth had cooled to its present state from an initially hot, largely molten, body. However, the discovery of radioactivity led to the realization that most of the heat currently being released through the Earth's surface must be derived from the breakdown of radioactive elements known to be present in significant quantities, principally uranium, potassium and thorium. Since heat is being continuously produced within the Earth, the present temperature distribution cannot be used to calculate its age. However, dating of rocks based on rates of radioactive decay showed that the oldest rocks were around 4000 million years old, and we now believe the age of the Earth and Solar System to be 4600 million years old (4.6Ga) based on meteorite ages.

Early history of the Earth

It is believed that the ancestral Earth grew by the accretion of particles from the solar dust cloud. As its mass increased, the resulting gravitational effect would have enabled it to sweep up most of the material in its vicinity. The energy produced by this accretion process increased the temperature of the interior to the point at which it was hot enough to melt iron, nickel and other similar heavy metals, which then drained down to form the core – the process illustrated by the meteorites, as explained above. At the same time, a primitive crust would have formed out of lighter silicate fractions. However, the volatile components that would have formed the earliest atmosphere would have escaped into space due to the high surface temperatures. As the Earth cooled, a new atmosphere would have been generated from comet impacts and volcanic exhalations, releasing water, nitrogen and carbon gases, and the so-called 'inert' gases mentioned earlier. With further cooling, the liquid metallic core gradually solidified from the centre outwards until now only the outer 2000km or so are still liquid.

At around 4.4Ga, the ancestral Earth is thought to have collided with another planetary body, known as 'Theia', about half the size of the Earth. This catastrophic event resulted in the destruction of Theia and the addition of much of its mass to the early Earth. The remainder became the circulating satellite body we now know as our Moon. It has been suggested that the Earth's core may have grown by around 20% as a result of the additional material, giving it sufficient gravitational pull to retain its atmosphere.

The early crust must have experienced considerable changes and additions of material during the period of intense meteorite bombardment that occurred between 4.1 and 3.8Ga, and was responsible for the prominent impact craters seen on the surface of the Moon (Fig. 2.6).

In the early days of the Earth–Moon system, the Moon would have been much closer and spinning much more rapidly. It has been calculated that this would have produced a greater tidal pull with four-hourly tides and up to 100 metres of tidal range. As well as the effects on the surface water, it is believed that the increased tidal pull would have induced cracking in a thin oceanic crust, enabling magmas to rise easily to the surface. Hot springs formed at these sites of vulcanicity may have encouraged the growth of early life.

The oldest known terrestrial minerals are zircon grains dated at

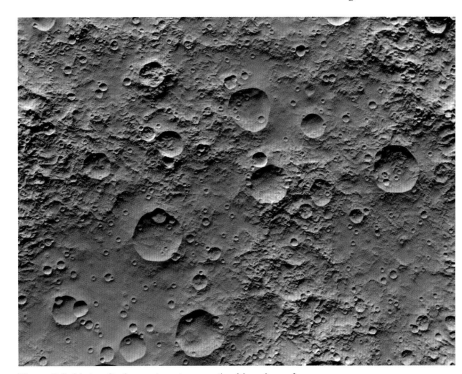

Figure 2.6 Meteorite impact craters on the Moon's surface
Image credit: © Shutterstock, HelenField.

*c.*4.4Ga, which indicate that granite-like material was available then. There is evidence from the early **Archaean** [1] of the existence of continental crust, modern-looking volcanics, and water-laid sediments by *c.*3.8Ga, indicating that, by around that time, processes similar to those associated with present-day plate tectonics were taking place. However, the higher heat flows in the Archaean Earth would probably have resulted in faster plate movements and smaller, and maybe thinner, plates.

There is evidence that early life existed by 3.5Ga [2] in the form of **stromatolites**, which still exist today. These are colonial bacteria that converted carbon dioxide from the early atmosphere into energy using sunlight, i.e. by **photosynthesis**, releasing oxygen as a by-product. For perhaps another 1000Ma, the oxygen released by these organisms was captured by the iron contained in the volcanic rocks of the ocean floor, forming layers of iron oxide; these built up to produce the **banded iron formation** so typical of the Archaean and Early **Proterozoic** Eons and the source of much of the modern supply of iron (*see* Fig. 5.1A, B). Once the available supply of free (uncombined) iron had been used up, towards the end of the Archaean, oxygen was available to enrich the atmosphere and increased in proportion to the point when oxygen-breathing life forms became possible.

Another important factor in the early environment was the influence of the igneous activity produced by the primitive plate-tectonic process, which was responsible for the generation of granitic magmas at subduction zones (*see* chapter 3). This granite formed the nucleus of the early continental crust, which, when uplifted and weathered, transported alkaline minerals into the oceans; these eventually neutralized the acidic sea water, making it a more hospitable environment for life.

By the beginning of the Proterozoic Eon, plate-tectonic processes appear to have become indistinguishable from those operating now, and substantial areas of relatively stable continental crust had been formed. These processes perform a vital function in redistributing and concentrating the elements, thereby producing economically important natural resources, as will be seen in the following chapter.

1 [*See* Appendix, Table A1 for the subdivisions of Geological Time.]

2 [It has been suggested that even the earliest known banded iron formation may also have been biogenic, putting the origin of life on Earth back to around 3.8Ga].

3 Redistribution and concentration of mineral resources

The abundance of the elements in the Earth as a whole bears no relationship to the likelihood of finding them on or near the surface. Many elements are extremely rare and are only found in minute quantities, measured in parts per million (ppm) in the commoner rocks. The reason why we are able to access useful natural resources of both minerals and rocks is that they have become concentrated by geological processes of one kind or another. This chapter examines the various processes by which such concentrations are achieved.

Plate tectonics

Plate tectonics is essentially a process of recycling material from deep within the Earth to the surface, and, in doing so, also transferring heat outwards. Slow convection currents within the solid mantle convey hot material from the interior close to the surface, where it melts to yield magmas with an initially basaltic composition. Cooled material is returned to the mantle at **subduction zones**, thus completing a vast circulatory system (Fig. 3.1). The process is driven by the temperature gradient between the core and the surface, which in turn produces a density gradient – warmer, less dense material rising upwards is balanced by cooler, denser material sinking downwards. This circulatory

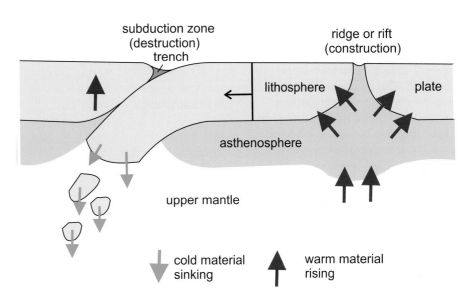

Figure 3.1 Earth's convection system
The lithosphere gains warm material from the asthenosphere at constructive plate boundaries (e.g. ocean ridges and continental rifts); during plate destruction, the lithosphere loses cold material at subduction zones (ocean trenches). The lithosphere plate moves towards the subduction zone.

movement takes place in the solid state at very low rates of only a few centimetres per year. The cooled surface layer, containing the crust and uppermost mantle, around 100km thick on average, is termed the **lithosphere**, and forms the tectonic plates whose relative movement provides the motor for most geological activity at the surface, such as the formation of mountain belts. The cool, strong lithosphere rests on a weak layer of

the upper mantle, termed the **asthenosphere**, and it is this layer from which the basaltic melts are derived.

The Earth's mantle is made of material with a composition similar to that of the chondritic meteorites described in the previous chapter. The Earth's crust grows by the addition of igneous rocks derived initially from the melting of this mantle material, either by extrusion on the surface by volcanoes, or by intrusion beneath it. Part of this activity takes the form of the injection of magma directly from the mantle at extensional **rift** zones, while part is derived from the release of secondary magmas produced indirectly as a result of the subduction of oceanic plates.

Ocean ridges and continental rifts

New ocean crust is created by extension of the oceanic lithosphere at ocean ridges (Fig. 3.2A) and also within continental lithosphere at extensional rifts (Fig.3.2B). This extensional process can often be related to the existence of a mantle '**hot spot**' beneath the rift. The African Rift valley is a well-known example of such a structure. The process can be described as follows: a zone of hot mantle first creates a topographic dome or ridge, which is then injected by magma rising along fissures within a central rift zone, as shown in Figure 3.2B. Continued extension results in the continental crust splitting and separating, to create new oceanic crust in the space between. In this way, continents split up and their fragments drift away as separate new continents – one of the fundamental processes of plate tectonics throughout geological time. The newly created continental margins are termed '**passive margins**' (Fig. 3.2C) since, in contrast with active margins (i.e. subduction zones), they lack significant tectonic or magmatic activity once they have migrated some distance from the original site of the rift.

Typical sedimentary rocks of a passive continental margin include clastic deposits (such as sandstones and siltstones) and **carbonates** (i.e. **limestones, dolomites,** etc.). Carbonates are generally confined to the continental shelf, whereas clastic sediments are also

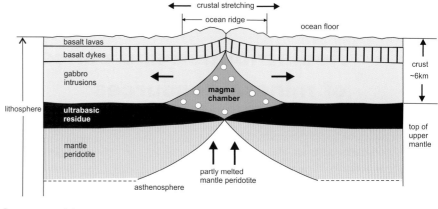

A ocean ridge

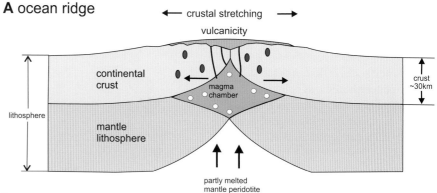

B continental rift

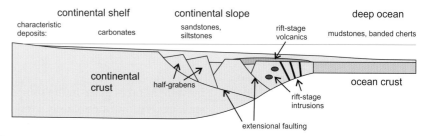

C continental passive margin

Figure 3.2 Ocean ridges, continental rifts and passive margins

A New oceanic crust is created at an ocean ridge by the addition of magma from the upper mantle. **B** The continental crust is stretched, thinned and uplifted above a mantle hot spot, from which magma intrusions and vulcanicity are generated. **C** The continental crust has separated to allow new oceanic crust to form while the stretched and faulted continental margin is depressed and covered in sediment: carbonates are characteristic sediments on the continental shelf, clastic deposits (e.g. sandstones) on the slope, and muds or cherts on the deep-ocean floor.

prominent on the continental slope, grading to fine-grained mudstones on the deep ocean floor, where bedded **cherts** are also typical (Fig. 3.2C).

Subduction zones
The process of subduction results in vulcanicity, either at the margin of a pre-existing continent, as in Figure 3.3, or along an island arc outboard of the continent. As the subducting slab descends into the asthenosphere, magmas, initially basaltic in composition, are produced by the partial melting of the asthenospheric mantle above the slab. This process is described in more detail below. The basalt magma rises through the overlying mantle lithosphere into the crust above to create a volcanic arc. This arc may be situated either on pre-existing continental crust or on oceanic crust.

Over a long period of time, a succession of volcanic arcs, together with the products of the intervening basins, may be added to the margin of a continent, which thus grows by accretion. Eventually another continental plate collides with it to form a broad **orogenic belt**. The continental crust has evolved in this way through geological time by successive additions of the igneous products of subduction-derived magmas.

Igneous processes
Almost all igneous magmas are generated initially from basaltic melts originating from the upper mantle. The wide variation in composition that they exhibit now is due to four separate processes: *partial melting*, *fractional crystallization*, *liquid immiscibility* and *crustal assimilation*. Each of these processes results in the

concentration of certain elements or rocks that may constitute a natural resource. To appreciate how these natural resources have been formed, it is necessary to understand how magmas are formed in the first place.

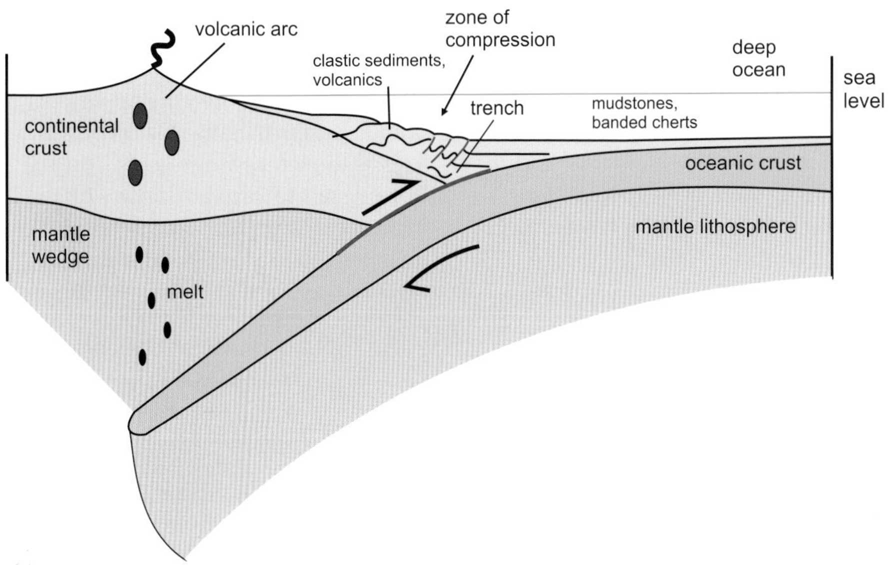

Generation of basalt magma
Basalt magmas are the result of melting of that part of the upper mantle known as the asthenosphere (see Fig. 3.1) and takes place at temperatures of over 1200°C at depths of between 50 and 200km. The composition of this part of the mantle is assumed to be a type of peridotite known as **lherzolite**, corresponding to that of the chondritic meteorites described in the previous chapter (table 3.1). Material with this composition is known from solid inclusions (**xenoliths**) brought up in basalt magmas, and from the results of melting experiments in which basalt melts have been produced from a peridotitic parent.

The basalt magma is produced by the process of **partial melting**, whereby those components of the source rock with the lowest melting point melt first, leaving a residue of components with higher melting points. In the case of basalt melts, the residues mainly consist of the minerals olivine, pyroxene and spinel, rich in elements such as magnesium, nickel and chromium, whereas the melts become enriched in alkaline oxides, silica and the so-called 'incompatible elements'.

The incompatible elements are of two types: those with larger ionic radii, termed **large-ion-lithophile elements (LILE)** such as rubidium, potassium and strontium, and those with higher ionic charges, the **high-field-strength**

Figure 3.3 Subduction at an active continental margin
Magmas derived from the melting of oceanic lithosphere, which is subducted beneath the continental margin at an ocean trench, ascend into the continental crust.

elements (HFSE) such as phosphorus and titanium; both are characteristics that make it more difficult for these elements to fit into the crystal structures of the common silicate minerals.

The conditions for this process to occur are confined to four specific types of location: ocean ridges, continental rifts, subduction zones and intra-plate (i.e. 'within-plate') hot spots. Elsewhere in the plate-tectonic network, beneath 'normal' lithosphere, the asthenosphere is not hot enough for melting to take place. However, melting is possible at ridges and continental rifts because the lithosphere has been extended and thinned, resulting in a decrease in pressure of the asthenospheric mantle, which in turn has depressed the melting point of peridotite.

Melting at intra-plate sites of volcanic activity is usually ascribed to the presence of upwellings of untypically hot mantle. These result in abnormally hot conditions in the lower part of the lithosphere, which has the effect of bringing melt-generating asthenosphere nearer the surface.

Melting at subduction zones
Melting here is due to a different process. As the oceanic crust of the subducting slab descends into the asthenosphere, at depths of between 60 and 90km, hydrated mineral assemblages such as **amphibolites** are converted into anhydrous rocks such as **eclogite**, releasing water and other volatiles, which pass upwards into the mantle wedge above the slab as shown in Figure 3.3. These in turn decrease the melting point of the solid asthenospheric mantle, which partially melts, producing basalt magma.

Chemical differences between magmas from different sources
Each of the above melting scenarios results in basaltic magmas that can be distinguished geochemically: typical ocean-ridge basalts, known as N-MORB ('normal' ocean-ridge basalts) contain rather more silica than continental rift basalts, which are usually more alkaline and contain less magnesium.

Table 3.1 gives the proportions of major oxides in N-MORB, continental-rift and island-arc basalts compared to the estimated mantle composition. The basalts show obvious increases in alumina, calcium oxide and the alkalis (sodium and potassium oxides), together with a very marked decrease in magnesium oxide. However, the differences in major oxides between the various types of basalt are not so marked. Alkali basalts are slightly enriched in iron and sodium and depleted in magnesium and calcium compared with MORB, whereas island-arc basalts are rather less enriched in iron and depleted in magnesium. The most striking differences are found in the incompatible elements, and so measuring the relatively small proportions of these (only a few parts per million) is an effective way of distinguishing between basalts of differing origins.

Subduction-zone magmas, although they include basalts, are typically more **andesitic** in composition and may also include more acid types such as **rhyolite**. This is due partly to the fact that the denser basaltic melts are inhibited from rising up to the upper layers of a thick continental crust and tend to form magma chambers at depth, from which more acid fractions of melt are produced. Another reason is that the hotter basalt magma is able to melt silica-rich crustal material (i.e. with a lower melting point) through which it passes, thus increasing the silica content of the melt. Such magmas are said to be more 'evolved', that is, their composition has moved further away from the original peridotite.

Fractional crystallization
The process of **fractional crystallization** is the opposite of partial melting. It takes place within magma chambers, either in the crust or in the uppermost mantle, where magma slowly crystallizes as the temperature cools. The first crystals to form are those with the highest melting point, and in basalt magma consist mainly of olivine and

major elements (oxides)		mantle	MORB	rift basalt	island-arc basalt
silica	SiO_2	44.3	50.5	47.2	50.8
magnesium oxide	MgO	41.6	8.1	5.7	5.7
iron (ferrous) oxide	FeO	8.3	8.9	13.4	11.2
alumina	Al_2O_3	2.4	16.9	14.8	16.2
calcium oxide	CaO	2.2	11.9	8.8	9.2
sodium oxide	Na_2O	0.2	2.6	3.6	2.8
potassium oxide	K_2O	0.0	0.1	1.3	0.9

Table 3.1 Mantle composition compared with various basalt types Proportions of the oxides of the major elements in weight per cent in average mantle compared with ocean-ridge, continental rift (alkali basalts) and island-arc basalts. Data source: Gill, R. 2010. *Igneous rocks and processes*, Wiley-Blackwell.

spinel, both rich in magnesium. These early crystal deposits are known collectively as the **cumulate**, and their release from the magma changes the composition of the remaining melt, depleting it in the constituents of the cumulate phases. The process is summarized diagrammatically in Figure 3.4. Cumulates, of which the ultrabasic rock **dunite** is a typical example, are important sources of economically valuable mineral deposits such as nickel, chromium, and elements of the platinum group. Many cumulate deposits show internal layering caused by gravitational settling of the heavier components within the individual layers before they solidify.

As fractionation proceeds, with progressive lowering of the temperature, the remaining melt fraction becomes more enriched in silica, alkalis and incompatible elements, eventually reaching a composition similar to rhyolite (i.e. granitic). At each stage in the process, liquid can be drawn off from the magma chamber to form a separate igneous body, whose composition, as determined by the composition of the melt at that stage, will be quite different from that of the original basalt.

The final products of the fractionation process take the form of aqueous fluids rich in incompatible elements that are the source of the important mineral veins that surround or cut granite intrusions; these are a valuable source of metallic ores such as those of tin, copper, lead and uranium. Aqueous fluids and gases are important products of volcanoes: the gases of the atmosphere are being continuously augmented by the emission of, principally, water vapour, carbon dioxide, and sulphur dioxide, as well as minor amounts of other gases such as chlorine and fluorine. Volcanic vents on the ocean ridges, the so-called '**black smokers**', are surrounded by rich deposits of metallic sulphide ores.

Liquid immiscibility

During the process of crystallization of a magma, certain liquid components may separate out from the main body of magma because they are incapable of mixing, rather like oil and water. Such liquids are termed **immiscible**. Sulphides, for example, form an immiscible liquid that may either sink to the base of a magma chamber, or be intruded separately into the host rock as veins, forming an important source of metallic ores.

Assimilation

A magma passing upwards through the crust has the potential to melt any portion of the surrounding host rock that has a lower melting temperature than the magma. Thus, basic magma will usually be hot enough to melt a silica-rich host rock such as a quartz sandstone, in which case it will become enriched with silica and possibly other minerals such as the alkalis, but basic rocks will not be assimilated. Once the quartz has been removed from a piece of host rock, the rock disintegrates, leaving the unassimilated components as xenoliths. Basic magmas that have assimilated sufficient material will thus become intermediate or even acid in composition.

Metamorphic processes

Metamorphism is the process in which the mineral assemblage of a solid rock is transformed into a different assemblage under the influence of heat and/ or pressure. Certain minerals are only

Figure 3.4 Fractional crystallization

Three stages in the evolution of a slowly cooling magma: in 1, the crystals with the highest melting point (e.g. olivine) settle to the bottom of the magma chamber first, followed successively in 2 and 3 by those with lower melting points (e.g. pyroxene and plagioclase respectively). The liquid composition changes as the various solid components are removed, and liquids drawn off at stages A, B and C therefore have different compositions.

liquid composition A liquid composition B liquid composition C

magma chamber

cumulate

1   2   3

1200°C   cooling   600°C

stable under a limited range of temperature and pressure, and if this range is exceeded, such minerals will break down and convert to new minerals that are stable under the changed conditions. Where such a transformation is accompanied by fluid transfer, the process is known as **metasomatism**.

Hydrothermal alteration

The role of water in transporting elements within the crust is very important in the formation of certain mineral resources; many ores of metals, such as copper and gold, are formed in this way. Both sea water (**brine**) and rainwater are commonly present at upper levels of the crust, circulating through cracks and pore spaces in the rock; water may also be derived from lower levels in the crust. Hydrous rocks tend to be unstable in the lower crust because of the higher temperature, causing water to be expelled from them; this is termed **dehydration**. The water that is driven off rises up to affect rocks in the upper part of the crust, and in its journey upwards may dissolve soluble elements and redeposit them elsewhere. Hot water, and particularly acid or alkaline hot water, is capable of dissolving many compounds, such as silicates, that have very low solubility under normal surface conditions. Mobile fluids formed in this way are termed **hydrothermal**, and the compositions of both the original dehydrated rocks and the new hydrated host rocks will naturally have been changed by this metasomatic process. Hydrothermal activity takes place under progressive regional metamorphism as the crust is warmed and thickened during the formation of an orogenic belt, and also at a more local level around igneous intrusions.

Deposition of minerals from hydrothermal fluids occurs in several ways. The solution may become unstable with falling temperature or pressure, or by the release of volatiles (**degassing**) causing the minerals to crystallize. Alternatively, the fluid may react chemically with the host rock, causing new minerals to form incorporating components from both the host rock and the hydrothermal fluid.

Examples of metasomatic processes

Skarns are concentrations of hydrothermally altered rocks formed around igneous intrusions such as granite, particularly where the intrusion is in contact with limestone. This is a consequence of the reduction of the solubility of certain metallic compounds when their host fluid passes from granite into a carbonate rock, causing them to be precipitated at their interface; this is an important source of metallic ores such as those of copper.

Kaolinization is the specific hydrothermal process of converting feldspar into kaolin, or **china clay**, in and around granite intrusions such as the Dartmoor and St Austell granites of SW England.

Sulphur-bearing hydrothermal fluids are also important in dissolving and transporting metallic elements such as iron, copper and zinc to form deposits of metallic sulphides.

Hydrous minerals such as **serpentine**, **chlorite**, **muscovite** and **amphibole** are commonly formed by the hydrothermal alteration of igneous rocks containing olivine, pyroxene or feldspar, and may form substantial deposits within or around them, or in faults or **shear zones**, often aided by deformation.

Weathering, erosion and deposition

Geological history records the continual creation of new land surfaces and their subsequent destruction. These destructive forces include the weathering, erosion and transport of rock material, and new, sedimentary, rocks are formed by the deposition of this material elsewhere on the land surface or in the sea. The redistribution of material caused by these processes results in significant concentrations of rock and mineral resources.

Weathering

This process changes the constituent particles of a rock in such a way that they are more easily broken down and worn away by erosion. The process may be mechanical, chemical or biological in nature, or some mixture of these.

Mechanical weathering may be caused by the repeated alternations of expansion and contraction due to extremes of temperature, found, for example, in desert regions, which produce weakening of the rock along the grain boundaries and eventual separation of the rock particles. Frost shattering is a similar process that works by the action of water expanding as it freezes within cracks in the rock.

Chemical weathering is caused by the chemical action of rainwater on rock minerals. Most rainwater is slightly acid due to dissolved atmospheric carbon dioxide. Acid water will dissolve some minerals, but only in minute quantities. However, a more important effect is to alter the chemical composition of

certain minerals, making them more susceptible to breakdown. For example, iron-rich silicates may be oxidized, producing a crust of red **haematite** on the surface and along cracks within the rock, and alumino-silicates such as feldspar break down to produce clay minerals. Many minerals react with water in a process known as **hydration**, which causes an expansion of the rock, thus facilitating weathering. Of the main rock-forming minerals, quartz is the least susceptible to chemical weathering, followed by muscovite mica, which explains why these two minerals are so common in sand. Both mechanical and chemical weathering can be caused or aided by organisms; thus tree roots may exploit cracks to force quite large blocks of rock apart, and lichens gradually dissolve the rock on which they grow.

Erosion

Rock material that has been loosened and disintegrated by weathering is transported elsewhere by the action of water, wind and ice. During heavy rain, sand-sized particles are washed downhill along channels worn by flowing water; strong winds transport fine dust or even sand; and in higher mountains with active glaciers, the ability of ice to transport quite large boulders is clearly evident.

Deposition

The rate of flow of a river from its source determines how effective it is in transporting rock particles. In places where the slope eases and the flow decreases, material is deposited as sediment, because the water is no longer able to carry it.

The smaller the size of the rock fragments (known as **clasts**) the more easily they can be carried along by the flow of the river, which causes a separation of the clasts according to their size, the smallest being transported furthest. Only the smallest particles of clay or mud are carried in suspension. This process is called **sorting** and has the effect of separating deposits into different sections, each characterized by a particular clast size; some parts by mud, some by sand and some by pebbles or boulders. This process is important in understanding how the different types of sedimentary rock are formed and separated from each other to form conglomerates, sandstones and mudstones, for example.

This sedimentation process continues towards the sea as the transported material is distributed in the estuaries of the major rivers, and beyond onto the continental shelf, continental slope, and ultimately onto the deep ocean floor, where generally only the finest material remains. It is these estuarine and marine deposits that are typically represented in the geological record and are an important source of industrial resources.

Deposits formed on land are rarely preserved in the geological record, except those formed in the relatively recent past or forming now. River and beach deposits may carry significant concentrations of certain ores; these are known as **placer deposits** and are important sources of gold and titanium. **Laterites** are soils enriched in iron, aluminium or nickel, depending on the composition of the bedrock. Glacial sands and gravels deposited during the later stages of the last Ice Age in northern Europe and North America are an important source of building materials.

Organic activity

All organisms need to ingest or react with certain elements in their immediate environment in order to survive. Obvious examples of this are the requirement of plants for carbon dioxide, animals for oxygen, and shell-bearing marine organisms for calcium. Organic activity is thus responsible for the large-scale concentrations of elements such as carbon and calcium in the form of coal and limestone deposits. However, many other elements are concentrated in this manner. Most plants and animals require small quantities of trace elements to survive. In addition to the obvious requirement for oxygen, water and carbon, a healthy human body requires small amounts of many other elements: e.g., nitrogen, calcium, sodium, potassium, phosphorus, magnesium, iron, fluorine, iodine and zinc. The uptake of trace elements by plants means that soils are thus progressively depleted in these elements.

Examples of large-scale natural resources that can be formed as a result of organic activity include coal, petroleum and natural gas from organic remains, limestone from animal shells, iron formation from bacteria, and phosphates from bird **guano**.

4   Types of ore deposit

Most metallic elements occur in the Earth's crust in the form of compounds such as oxides, silicates, chlorides or carbonates; only a few, such as gold and silver, occur in the uncombined state (i.e. as 'native' elements). Some metals (e.g. aluminium, iron, titanium and manganese) are present in significant quantities in the igneous rocks that are the original source of the metals. All the other metals are present in minute quantities, measured in parts per million (ppm). Thus to form an economically viable resource, the metallic ore must be concentrated many hundreds or thousands of times. In many cases there must also be some kind of transportation mechanism to transfer the metal to a new site where the ore can be concentrated. Ore geologists recognize a number of different types of ore deposit, depending on the nature of the location, presumed transport mechanism and origin, of which the more important ones are now described.

Deposits related to igneous activity

Although all the metallic elements are ultimately derived from igneous magmas, it is convenient to distinguish between those ore deposits that are directly associated with igneous bodies, and those where the ore material has been transported from its origin and deposited elsewhere. The four main types of deposit in the former category are as follows.

Volcanic-hosted massive sulphide ore (VHMS) deposits

This type of deposit is mainly exploited for copper and zinc, but also contains iron, lead, gold, and silver, together with many trace metals as by-products, including cobalt, tin and manganese. The present-day setting for the formation of these deposits is the oceanic **black smokers**, which are submarine volcanic hydrothermal vents. These form black, chimney-like structures that rise from the sea bed, typically along active ocean ridges, and pour out a cloud of black 'smoke' containing metallic sulphides that precipitate out in contact with the cold water. Oceanic examples cannot at present be exploited, and land-based examples are mostly situated in former extensional rifts and **back-arc basins** that have been thrust onto either a volcanic arc or the continental margin. These deposits typically lie above an acid volcanic sequence, and are often overlain by more basic volcanics.

It is believed that the VHMS deposits form by the extraction of certain elements, including metals and sulphur, from the hot volcanic rock by means of the hydrothermal circulation of saline ocean water that has been heated from the volcanic source. The saline water cools as it passes through the surface rocks, depositing the metal sulphides in a kind of apron around the volcanics.

The ore body may be shaped like a bowl, when the ore has been deposited in a hollow in the ocean floor, or a mound, reflecting the form of the original volcanic vent. The ore body typically consists of an internal **stockwork** of altered and veined volcanic rock, surrounded by the main body of massive sulphide ore, which is flanked in turn by a **stratiform** (bedding-parallel) sulphide layer. The stockwork consists of metasomatically altered volcanics cut by sulphide veins containing mainly **chalcopyrite** (copper-iron sulphide), **pyrite** and **pyrrhotite** (both iron sulphides). The massive sulphide body, which may be several tens of metres thick and hundreds of metres wide, typically contains pyrite, **sphalerite** (zinc-iron sulphide), **galena** (lead sulphide), **haematite** (iron oxide) and **barite** (barium sulphate). The outer apron is similar to the **SEDEX** ores described below, and is generally rich in manganese, barium, haematite and chert (silica). These ore bodies typically contain more than 90% of sulphide ore.

Porphyry copper ore deposits

This type of ore body is the most important source of copper at the present day, and is found in large numbers along currently active, or recent, subduction zones such as those along the western coastal zone of the Americas and around the Pacific margin. Chile is a particularly important source region. These bodies are also a source of gold and molybdenum. The ore is of a low grade (less than 1%

copper) but this is offset by the large volumes of the deposits, which are extracted by **opencast** (surface) mining.

The ore body (Fig. 4.1A), consisting mainly of sulphides, is formed at a relatively high level in the crust, in the form of a stockwork of veined host rock. The ore complex is centred around an intrusive body known as a **stock** acting as the feeder to a volcanic vent (which may or may not be exposed) and is accompanied by numerous intrusive dykes. Both the dykes and the stock are composed of **porphyritic** igneous rock with an intermediate composition, typically **diorite**, indicating that the magma source was relatively evolved (i.e. fractionated). The tectonic setting of these bodies means that they will ultimately be involved in a continental collision zone, and either be destroyed or difficult to recognize in the geological record, which explains the fact that most of the known examples were formed in the Cenozoic Era.

Iron oxide-copper-gold ore (IOCG) deposits

This type of deposit is an important source of copper, gold and uranium, and usually occurs within an iron-oxide rich host material. It typically provides between 0.25% and 5% copper and 0.1–0.3ppm of gold, along with trace amounts of uranium, bismuth and rare-earth elements. Some of these deposits are extremely large and profitable, for example the Olympic Dam deposit in South Australia. The main ores are pyrite and chalcopyrite; however, the copper ore may be metasomatically altered to **malachite** (copper carbonate), **cuprite** (copper oxide) and native copper. Rare-earth elements may be associated with phosphates.

The ore occurs within and around the margins of large granodioritic or dioritic intrusions, often in regional-scale 'ore provinces'. They occupy a similar tectonic setting to the **porphyry copper** deposits, and originate via the same magmatic-hydrothermal process, but they represent a deeper crustal level (Fig. 4.1A). Because of their deeper position, they exhibit greater metamorphic effects, ranging from **greenschist** to **amphibolite facies**, and the ore is often found in or near deep-crustal shear zones. Most of the deposits occur in Precambrian basement rocks and are often associated with iron-rich host rocks such as banded iron formation.

Figure 4.1B shows how the metallic ores of a typical

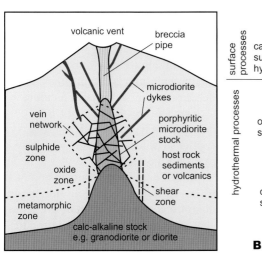

A Magmatic-hydrothermal system

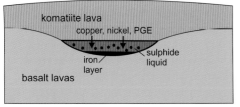

C Magmatic nickel-copper-PGE deposit

B Idealised zonation of magmatic-hydrothermal ores

Figure 4.1 Ore deposits formed by magmatic processes

A Diagrammatic representation of a magmatic-hydrothermal system: the upper part corresponds to the porphyry-copper type deposit, the lower to the iron ore-copper-gold type. The zone of sulphide mineralization occurs in the vein network within and around the porphyry stock, the breccia pipe and the associated micro-diorite dykes, and in the lower, metamorphic part of the system, in shear zones. **B** Vertical zonation of ore types in a magmatic-hydrothermal system. **C** A typical magmatic nickel-copper-PGE ore deposit: the ores precipitate from a pool of immiscible sulphide liquid that accumulates in a hollow at the base of a komatiitic lava flow.

magmatic-hydrothermal system vary upwards from an oxide-silicate zone through a zone of oxides and sulphides to an upper zone, dominated by surface processes, of carbonates, sulphates and hydroxides. Copper and gold may be found throughout, and in the weathered zone may occur as native metals, whereas lead and zinc are preferentially concentrated in the topmost layer.

Magmatic nickel-copper-platinum group ore deposits

These deposits are associated with large basic to ultrabasic intrusive sills or lavas, often layered, or their stock-like feeder bodies. The magmas originate in the upper mantle; the sulphur may represent the original content of the magma, or it may have been introduced by assimilation from the host rocks. An immiscible sulphide melt forms, dissolving nickel, copper, iron and **platinum-group elements (PGE)** from the magma, and sinks to the base of the magma chamber, where it forms an ore body, which may be in the form of a concordant sheet; alternatively, the ore may migrate into local breccia zones or faults.

The ore consists of iron-nickel-copper sulphides with trace amounts of PGE metals in addition. Sulphide ores include **pentlandite** (nickel-iron sulphide), chalcopyrite and pyrrhotite. Some deposits contain silver, selenium and tellurium in addition to the PGE.

There are two types of ore deposit, depending on sulphur content: one in which the sulphur has been largely introduced, where nickel is the principal economic mineral; and the other where the sulphur is from the original magma, and PGE are the more

important economic minerals. In the PGE deposits, the ore is usually more dispersed within the host intrusion.

Most deposits of this type (e.g. the Bushveld Complex in South Africa) occur in Archaean or Early Proterozoic rocks. However, one of the largest is the Noril'sk-Talnakh deposit in Siberia, which is of Permo-Triassic age. All of these produce very large volumes of ore.

Recognized as a distinct subclass of these magmatic ores are the *Kambalda-type komatiitic nickel ore deposits*, named after the type example of Kambalda, in the Yilgarn region of Western Australia. This type of ore body (Fig. 4.1C) is restricted to komatiitic volcanics, which are characteristic of the Archaean. **Komatiites** are ultrabasic lavas formed from exceptionally high-temperature melting of the mantle source. The ore has been deposited into hollows or channels on the contemporary land surface, and then overlain by the ultrabasic lava. The ore body consists typically of a massive sulphide layer containing over 90% ore, overlain by a zone of mixed host rock and sulphide, grading up into a cumulate layer at the base of the lava sequence.

Ore bodies of the Kambalda type are usually metamorphosed to greenschist or amphibolite facies, and often subsequently retrogressed. The ore may become locally mobilized without changing the mineralogy. Weathered zones contain nickel carbonates and sulphates.

Ore deposits formed by surface processes

Several important types of ore deposit are formed through surface or near-surface processes such as sedimentary

deposition, surface weathering or the circulation of aqueous fluids. These processes redistribute and concentrate ore that has been originally produced from magmatic sources. Five specific categories are recognized by ore geologists.

Sedimentary-exhalative (SEDEX) ore deposits

This type of deposit is basically stratiform, that is, it forms a layer or layers parallel to the host strata. In that respect, it is similar to the outer apron of the VHMS ore bodies already described, although in the SEDEX case there is no obvious volcanic source. These deposits are believed to have originated from the release of saline fluids from the sedimentary pile as it is compressed and dehydrated during burial or tectonic activity. The ores found in SEDEX deposits are typically lead, zinc and copper sulphides, together with barite (barium sulphate).

SEDEX deposits that are hosted within carbonate beds (limestones and dolomites) are recognized by ore geologists as a special sub-type, referred to as 'carbonate-hosted lead-zinc deposits' or 'Mississippi Valley type deposits'. In this type of deposit (Fig.4.2A), the ore may be partly stratiform and partly in the form of discordant veins, which, although confined to a particular limestone formation, have obviously been introduced from the surrounding rock. The process is thought to have occurred at relatively low temperatures via brine (i.e. saline) solutions, which are capable of dissolving the lead (along with zinc and sulphur) from large volumes of sedimentary rock (e.g. clays and shales), and concentrating it in nearby permeable carbonate rocks.

Laterites

These are tropical weathering deposits formed in hot, wet climates where soluble minerals containing alkalis, magnesium, calcium and silica are leached from the soil and washed away during the wet season, leaving a residue of insoluble iron or aluminium hydroxides (Fig. 4.2B). Iron-rich laterites (e.g. **limonite**) form above basic igneous rocks and aluminium-rich laterites (**bauxite**) above iron-poor granitic or limestone bedrock.

Lateritic nickel ore deposits occur in the weathered zone above ultrabasic rocks rich in olivine, such as peridotite, dunite or komatiite. Such laterites are enriched in **goethite** – an iron hydroxide containing 1–2% of nickel. Beneath this layer is a silicate layer consisting of magnesium-depleted **serpentinite** that may contain over 1% of nickel. Because this type of deposit is essentially a surface rock, it can extend as a sheet for many kilometres and can be mined by opencast methods. Although the ore is relatively low-grade, the deposits can yield very large quantities of ore.

In addition to the metals listed above, the insoluble oxide residues may also contain small quantities of zirconium, titanium, tin and manganese.

Bog-iron ore deposits

A common type of iron-ore deposit forms at the base of a soil layer beneath a bog or marsh, where the groundwater is particularly acid (Fig. 4.2C). Iron is leached from the soil in the form of iron carbonate, and can be converted to insoluble limonite in a layer above the bedrock by oxygen-rich groundwater from the bedrock. In a similar type of system, limonite may be deposited on the floor of a lake by groundwater containing the dissolved iron carbonate (Fig. 4.2D). Both these types of iron-ore deposit have been widely worked in the past.

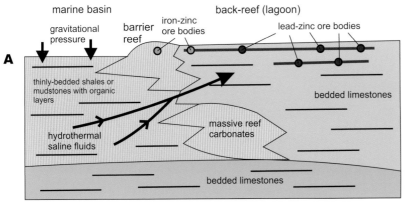

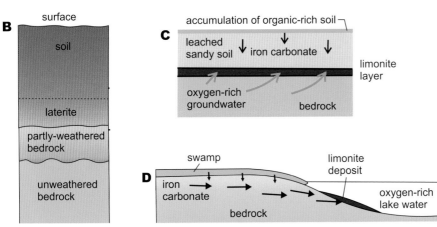

Figure 4.2 Ore deposits formed by surface processes
A Geological setting of a typical carbonate-hosted lead-zinc deposit: ore-bearing fluids are squeezed out of the marine basin sediments and move up into the back-reef limestones, where the ore is deposited in cracks and cavities in the upper layers.
B Laterite deposit: prolonged weathering of tropical soils causes soluble minerals to be leached out of the soil, leaving an insoluble layer of iron or aluminium oxide. **C** Acid groundwater in an organic-rich marshy environment leaches iron from the soil as a carbonate, which is oxidized by groundwater from the bedrock to form limonite. **D** At a lake margin, groundwater containing dissolved iron carbonate flows into the lake, where the iron is oxidized by the lake water to form a deposit of limonite on the lake bed.

Unconformity-related uranium ore deposits

These deposits are the richest source of uranium ore. They occur at, or close to, major unconformities between a quartz-rich sandstone cover and a crystalline, usually metamorphic, basement. They are typically of Proterozoic age, as are the two largest: the Athabasca Basin in northern Saskatchewan, Canada, and the McArthur Basin in the Northern Territories of Australia. The primary ore minerals are **uraninite** and **pitchblende**, both uranium oxides.

As uranium is incompatible in silicate melts, and very soluble, it can be easily dissolved, transported and redeposited by groundwater, leading to secondary deposits. The original source of the uranium would have been from late-stage aqueous solutions derived from highly fractionated granitic magmas.

Placer deposits

These are secondary deposits that have been laid down and concentrated by a sedimentary process, either in rivers or lakes, or on a beach. A large number of heavy minerals can be deposited in this way, including gold, zircon (zirconium silicate), **rutile** (titanium dioxide), **ilmenite** (iron-titanium oxide), and gemstones such as diamond and sapphire. Heavy-mineral sands are deposits formed in ancient beach sand-dune systems, which have either been uplifted above sea level or been buried beneath younger deposits. Sorting of the sands by wave action, perhaps assisted by severe storm events, will have removed the lighter sand, leaving the heavier minerals concentrated in a single layer. Heavy mineral sands are an important source of tin, titanium and zirconium, as well as the less common thorium and the rare-earth elements.

Gold placer deposits are found in modern river systems, where the gold is extracted by '**panning**', often by amateur prospectors.

5 Metallic mineral resources – I

Of the 103 elements in the Periodic Table, the great majority are regarded by chemists as metals although, to the layman, many of them might not look like the more familiar metals such as iron or aluminium. Going from left to right along the Periodic Table (see Fig. 2.2), the first column, beneath hydrogen, consists of the **alkali metals**, which include sodium and potassium; the second column consists of the **alkaline earth metals**, which include calcium and magnesium; to its right are four long rows of the **transition metals**, which include all the familiar metals such as iron, copper and zinc, together with the **post-transition metals**, which include aluminium and lead, but still look like 'normal' metals. The right-hand end of the Table consists of the non-metallic elements. The more common transition and post-transition metals are described in this chapter, the remainder in the following chapter together with the alkalis and alkaline earths.

Iron

Iron (Fe) (atomic number 26) is one of the most abundant elements in the Earth, forming about 85% of the core and occurring in most igneous rocks. The average ocean-ridge basalt contains about 9% of iron oxide and the average granite about 2%.

History of use

Iron has been used since at least about 1500BCE when its value in making tools and weapons was first discovered and gave its name to the 'Iron Age'. Even before that, iron was used occasionally for ornamental purposes. Iron is now used mainly in the form of **steel**, in which a small proportion of carbon (less than 2%) is added to iron to create the material that is so essential to the modern world. Certain trace elements, such as vanadium, may be added to steel to give it additional strength. Chromium, zinc and tin have all been used as coatings to prevent surface rusting of steel.

Occurrence and origin

Iron ores usually occur in the form of oxides, either **magnetite** or haematite, but also as hydroxides (e.g. limonite) or sulphides (e.g. pyrite). Haematite is familiar to everyone in the form of rust. Magnetite is, of course, magnetic, and its properties as a compass have been exploited by navigators since the twelfth century. Magnetite ore deposits, or even basic igneous rocks containing a high proportion of magnetite, can be easily detected with an ordinary compass. Some types of deposit are formed by hydrothermal activity, such as magnetite-bearing skarns around granite intrusions, and also as cumulate deposits in basic intrusions.

Iron ores are found in most of the types of ore deposit described in the previous chapter, and various of these types have been exploited in the past, particularly **bog-iron** ore (see Fig. 4.2C), which is an iron hydroxide; however, most present-day worked iron deposits such as the one illustrated in Figure 5.1A are in the form of banded iron formation (Fig. 5.1B). This is a type of organic iron-oxide deposit formed during the early Precambrian by bacteria as a waste product arising from their photosynthesis – the conversion of carbon dioxide and sunlight into sugars. This process releases oxygen, which combines with iron on the sea floor to form haematite or iron silicate; however, the haematite is usually converted by metamorphism into magnetite ore. The iron oxide is typically interbanded with quartz or iron silicate, as shown in Figure 5.1B. The role of these bacteria in the evolution of the early atmosphere is discussed in chapter 2.

Resources

The top five sources of mined ore at present are China, Australia, Brazil, India and Russia (table 5.1A). World production of unprocessed ore in 2013 was over 3 billion tonnes and world consumption is estimated to be growing at a rate of 10% annually. Available resources of iron are vast, in theory, and much is recycled. Although, at present rates of consumption, many of the present sources that are economically viable could become exhausted within this century, the future supply is not in danger. The reason for the recent growth in ore production has been

Figure 5.1 Iron

A The Carajás iron ore mine, northern Brazil, one of the world's largest opencast mines. Image credit: Jacquesd Jangoux. **B** 2.7 billion year-old banded-iron formation from Temagami, Ontario, Canada; the dark bands are magnetite, the red are jasper (iron silicate). Image credit: © Shutterstock, Paulo Afonso.

the steep increase in steel usage among the industrial or industrializing nations, especially China.

Aluminium

Aluminium (Al) (13), spelt 'aluminum' in North America, is one of the post-transition group of metals. It is the most abundant metal in the Earth's crust, making up around 8% by weight. It is also one of the most useful, as it is light, durable, soft and easy to work, and it resists corrosion by forming a hard, thin protective coating of aluminium oxide. An environmentally important quality is that it is very easy to recycle.

History of use

Aluminium salts were used by the Greeks and Romans for various purposes, but the metal itself was not isolated until 1827, by Friedrich Wöhler. His method of extraction using potassium was expensive, and the element was not used widely until the electrolytic process was made possible by the commercial generation of electricity in the 1880s. Hydroelectric power has been used in the production of aluminium in the Fort William area of Scotland since 1907, initially at Kinlochleven, and currently on the slopes of Ben Nevis.

The expansion of the aeroplane industry during the First World War demanded large quantities of aluminium in order to produce light but strong airframes. Since that time, its use has been extended to the construction of ships, cars, and trains, and many other familiar objects, such as drinks containers, household utensils and road signs. Like copper, it is also a good conductor of electricity and because it is lighter, is substituted for copper in overhead transmission cables. Its strength is increased by alloying with copper, zinc or magnesium.

Occurrence and origin

Most aluminium is produced from bauxite ore, much of which is in the form of gibbsite, an aluminium hydroxide, and is formed in tropical climates

iron ore production	million tonnes	steel production	million tonnes
China	1,450	China	823
Australia	609	EU	169
Brazil	317	Japan	111
India	150	USA	88
Russia	105	India	87

A. Iron and steel

country	amount in million tonnes
China	20.3
Russia	3.85
Canada	2.97
USA	2.07
Australia	1.86
UAR	1.82
India	1.7
Brazil	1.440

B. Aluminium

country	amount in million tonnes
Chile	5.78
China	1.6
Peru	1.38
USA	1.25
Australia	0.99
Congo	0.97
Russia	0.83
Zambia	0.76
Canada	0.632
Indonesia	0.504

C. Copper

country	amount in kilotonnes
Philippines	424
Russia	255
Indonesia	248
Australia	246
Canada	205
Brazil	139
New Caledonia	132
China	93.3
Columbia	76
Cuba	71

D. Nickel

country	amount in kilotonnes
Congo	51
China	7
Canada	6.63
Russia	6.3
Australia	5.88
Cuba	4.9
Zambia	4.2
Brazil	3.9
New Caledonia	2.62

E. Cobalt

Table 5.1 Metal production by main producing countries – I
Data sources: US Geological Survey (USGS), A, 2015; B–E, 2014.

varieties of corundum: ruby contains added chromium, and sapphire has added iron. All these minerals are distinguished by their extreme hardness.

The aluminium production process requires very large amounts of electricity (averaging around 15 kilowatt-hours per kilogram) and aluminium smelters are often situated near cheap sources of electricity: for example, using hydro-electric power in Quebec, Iceland, Norway and Scotland.

Resources

Total world production in 2012 was 44.4 million tonnes. The five top producing countries were China, Russia, Canada, USA and Australia (table 5.1B). Because of its abundance in the crust, the resources of aluminium are theoretically inexhaustible, and the easily exploitable bauxite could be replaced by other sources. However, the limiting factor is the cost of electricity, which would undoubtedly determine future production policy, although aluminium is easy to recycle and little is wasted in the process. Europe already recycles 42% of drink cans and 85% of construction material, and this practice could easily be extended.

Copper

Copper (Cu) (29) is one of the lighter metals in the Periodic Table, and one of the best known ones in everyday use.

History of use

Unlike iron and aluminium, copper can be found in the 'native' state, that is, uncombined with oxygen or other elements. For this reason, like gold and silver, it can be easily extracted in its native state from its host rock

by surface weathering of bedrock that is deficient in iron and silica. The process is similar to the one described in the previous chapter for lateritic nickel ores (e.g. see Fig. 4.2B), except that the bedrock is granitic (e.g. syenite) or limestone rather than ultrabasic, and relatively poor in iron. There are large deposits of bauxite in Australia, Brazil, China, India and Russia.

Aluminium is also found naturally in the form of oxides and silicates. The feldspar minerals, which are aluminasilicates of potassium, sodium and calcium, are the most abundant group of minerals in the Earth's crust. Aluminium oxide is found naturally as corundum, which is present in small quantities in many igneous rocks. It also combines with magnesium, iron and chromium to form the minerals spinel and chromite, and with beryllium and silica to form beryl. The gemstones ruby (Fig. 5.2A) and sapphire are coloured

Figure 5.2 Aluminium and copper minerals
A Ruby, a gemstone form of corundum (aluminium oxide). Image credit: © Shutterstock, Cherniga Maksym. **B** Green coating of malachite on the copper roof of the Church of St. Francis of Assisi in Prague. Image credit: © Shutterstock, Vitaly Titov and Maria Sidelnikova. **C** Malachite. Image credit: © Shutterstock, Zelenskaya. **D** Azurite. Image credit: © Shutterstock, Miriam Doerr.

by crushing, and therefore was widely used by several early civilizations for ornaments, coins and utensils; it was first used at least 10,000 years ago. It was eventually realized that by alloying it with tin, a much stronger material, **bronze**, resulted which could be used to make weapons – thus the Bronze Age, which archaeologists date from 3000 to 2000 BCE. Only when it was discovered that iron could be released from its oxide by smelting, was bronze superseded by the much stronger iron.

Bronze was much employed by the Romans for coins, instruments and jewellery. At that time much of the copper came from the island of Cyprus, giving copper the Latin name '*cuprum*' (thus, Cu). Bronze is still used widely, mainly for ornamental purposes. Another common alloy is **brass**, in which copper is combined with zinc, and which has also been used since Roman times.

The most important use of copper today takes advantage of its qualities as an excellent conductor of both heat and electricity. It is soft, flexible and ductile, making it ideal for electric wiring, which accounts for over half of the copper used now. It is also much employed in the manufacture of electric motors and electronic equipment. Copper also features in architecture as roofing material because of its resistance to corrosion and its attractive colour. Exposure to the weather over a period of years produces an outer coating of green copper carbonate, termed **verdigris**, which protects it from further weathering, and is a distinctive feature of the more prominent buildings in many ancient cities (Fig. 5.2B).

Small amounts of copper are essential to all living organisms to enable their cell structure to function. Despite this fact, it is also antibiotic, which in the past prompted its use on ships' hulls to protect them from barnacles. At the present time, its anti-bacterial qualities are being increasingly employed in the manufacture of hospital equipment. Copper tubing is much used in plumbing for the same reason.

Occurrence and origin

Copper is relatively abundant, making up from 60 to 70ppm (parts per million) of the Earth's crust, and is present in significant quantities (from tens to several hundred ppm) in basic igneous rocks. It occurs in the form of native copper, oxides (**cuprite**), sulphides (**chalcopyrite**) and carbonates (**malachite** and **azurite** – Fig. 5.2C, D). At present, most copper production comes from large opencast mines in copper sulphide deposits containing 1% or less of copper. Although copper, like iron, is found in many of the different types of deposit already described, most of the major exploited deposits are of the porphyry copper type, and are produced as a late-stage hydrothermal product of granitic intrusions (see Fig. 4.1A).

Resources

Total world production in 2013 totalled 18.3 million tonnes. The top five copper producing countries are: Chile, China, Peru, USA and Australia (table 5.1C). Due to its abundance in the crust, copper resources are potentially vast, but those that are economically viable at present are estimated to last for between 25 and 60 years at present rates of consumption. Much copper can be, and is, recycled.

Nickel

Nickel (Ni) (28) is a hard but ductile silvery-white metal which easily reacts with air to form a protective coating of nickel oxide. Most of Earth's nickel resides in the core as an iron-nickel alloy. It is also found in nickel-iron meteorites, which were probably the source of the iron-nickel alloy used in prehistoric times to make tools. In the crust, it occurs as nickel-bearing compounds of iron or magnesium.

History of use

Nickel is known alloyed with iron from about 3500BCE but was only isolated in 1751 by Axel Cronstedt. It was used until recently, mainly for coins and as an alloy in corrosion-resistant steel, i.e. **stainless steel**. Now, its main value is in the manufacture of specialist alloys that have corrosion-resistant or magnetic properties. Small quantities of nickel are an essential requirement for the human body but excessive amounts are poisonous. Contact with the skin can cause irritation, and its use in coins has been abandoned in many countries.

Occurrence and origin

Nickel is relatively abundant in the crust (*c*.80–200ppm) and makes up around 50ppm of the average oceanic basalt (but several hundred ppm in ultrabasic rocks). It is found in many types of ore deposit, often associated with copper. Most of the world's nickel resources are in the form of **pentlandite** (nickel-iron sulphide) in magmatic nickel-copper-PGE deposits (*see* Fig. 4.1C) such as those of Noril'sk-Talnakh in Russia, Sudbury, Ontario, Voisey Bay, Labrador, and the Kambalda-type deposits of Australia. However, most mined ore presently comes from laterites in the form of either nickel-bearing limonite (iron-nickel hydroxide) or **garnierite** (magnesium-nickel silicate)).

Resources

Total world production was estimated at 2.2 million tonnes in 2012. The top five producing countries are: the Philippines, Indonesia, Russia, Australia, and Canada (table 5.1D) and reserves are estimated at 74 million tonnes – the biggest being in New Caledonia and Australia. However, nickel can be easily recovered and re-used many times, and it is one of the most recycled metals.

Cobalt

Cobalt (Co) (27) is a hard, lustrous silvery-grey metal whose main industrial application is in steel alloys.

country	amount in kilotonnes
China	104
USA	60.4
Chile	35.1

A. Molybdenum

country	amount in kilotonnes
China	110
Indonesia	41
Peru	26.1
Bolivia	19.7
Burma	11

D. Tin

country	amount in kilotonnes
Australia	605
South Africa	380
China	140
Indonesia	120

G. Zirconium

country	amount in kilotonnes
China	2,900
Australia	711
USA	340
Peru	266
Mexico	210

B. Lead

country	amount in kilotonnes
China	64
Russia	3
Canada	2.19
Bolivia	1.27

E. Tungsten

country	amount in kilotonnes
South Africa	3,600
Australia	3,080
China	2,900
Gabon	1,650
Brazil	1,330

H. Manganese

country	amount in kilotonnes
South Africa	11,000
Kazakhstan	4,000
India	3,900

I. Chromium

country	amount in million tonnes
China	4.9
Australia	1.51
Peru	1.28
India	0.758
USA	0.738
Mexico	0.66

C. Zinc

country	amount in kilotonnes
China	80
Russia	44
Japan	40
Kazakhstan	25
Ukraine	10

F. Titanium

country	amount in kilotonnes
China	39
South Africa	19.5
Russia	15

J. Vanadium

Table 5.2 Metal production by main producing countries – II
Data sources: US Geological Survey (USGS): A, C–J, 2014; B, 2015.

History of use

Cobalt compounds have been exploited as pigments since ancient times: both the Egyptians and the Persians used cobalt blue for colouring glass and ceramics in the second millennium BCE, and the Chinese used it for their famous blue-patterned pottery. Until recently it gave the distinctive blue colour to medicine bottles. It was first isolated and named by Georg Brandt in 1735. Its main modern application is in steel alloys because of its high strength, wear-resistance and magnetic properties. Its silicate and aluminate compounds are still used as pigments, and its isotope, cobalt-60, is an important radioactive source in medical procedures. It is an essential dietary mineral, required for vitamin B12, and its deficiency leads to the inability of blood cells to carry oxygen.

Occurrence and origin

The crustal abundance of cobalt is about 25ppm, and it is present in proportions of between 10ppm and 50ppm in oceanic basalts. It occurs naturally in the form of oxide, hydroxide, sulphide and arsenide ores, associated mainly with nickel and copper, and is present in small amounts in most soils. The principal mined ore is cobaltite (cobalt-arsenic sulphide) which is produced as a by-product of copper-nickel mining from magmatic copper-nickel-platinum type deposits (Fig. 4.1A). It can also be sourced from sedimentary weathering products in the form of hydroxides and carbonates.

Resources

World production in 2012 totalled 103 kilotonnes, the main producing country being the Democratic Republic of Congo, where Mukondo Mountain mine produces about 40% of the entire world supply. Other major producers are China, Canada, Russia, and Australia (table 5.1E). World reserves are considerable – over 7 million tonnes in the top ten producing countries; also a considerable proportion of the cobalt used is recycled.

Molybdenum

Molybdenum (Mo) (42) is used mainly for making tough steel alloys. During the First World War, when Britain introduced the tank, it was discovered that only 25mm (1 inch) of molybdenum steel could give better protection against direct shell hits than the 75mm (3 inch) manganese steel plating that was first introduced!

History of use

Molybdenum was isolated from the mineral **molybdenite** (Fig. 5.3A) in 1781 by Peter Hjelm but only found an industrial application in the early nineteenth century. Now, it is the basis of high-temperature steel alloys for the automobile and aircraft industries, high-speed drilling machinery, and heating elements. Molybdenum steel can require up to 8% molybdenum.

Molybdenum is essential in trace quantities for all plant and animal life; it is a component of the enzyme nitrogenase, which allows the leguminous plants to convert nitrogen into nitrates. It only became accessible in the Precambrian oceans once oxygen became available in sufficient quantities in the Proterozoic, enabling its insoluble sulphides to be converted into soluble oxides. This allowed the eukaryotic cell, the basis of all higher life forms, to evolve.

Occurrence and origin

The crustal abundance of molybdenum is 1.1–1.2ppm. It occurs mainly in the form of its sulphide ore molybdenite, a soft, grey mineral, which looks so similar to graphite that it was formerly used as a substitute for it – e.g. to make the 'lead' in pencils!

Most molybdenite is recovered as a by-product of copper mining and is found in the lower zones of porphyry-copper systems (*see* Fig. 4.1A, B); the mineralization of the Shap granite in Cumbria, NW England, is an example.

Resources

World production of molybdenum in 2012 totalled 259 kilotonnes, mainly from China, the USA, and Chile (table 5.2A). World reserves were estimated at 11 million tonnes.

Lead

Lead (Pb) (82) is one of the heavier metals in the post-transition group. It has been widely used since Roman times because it is readily available and, like aluminium, is soft and easily worked: it can be rolled into sheets, fashioned into pipes, and welded. It melts at only 327°C on a simple wood fire. The Latin word for lead is *plumbum*, hence the symbol 'Pb'. The word 'plumbing' comes from this source.

Lead is a bright, silvery metal but rapidly tarnishes to a dull grey colour by reacting with rainwater and air to form an insoluble coating of lead carbonate or sulphate, which protects it from further corrosion. Because it is a radioactive decay product of uranium and thorium, measurement of its isotopic composition has given geologists a useful method of dating rocks.

History of use

Lead compounds used as cosmetics have been found in ancient Egyptian tombs, and the Romans mined lead ore extensively in many parts of Europe, including Britain, mainly for its silver

Figure 5.3A Molybdenite
Image credit: © Shutterstock, Bonchan.

content, using the lead for water pipes in their plumbing systems. They also used it to sweeten wine and to make drinking cups, not realizing that they were being slowly poisoned by it! More recently, lead has been deployed for roofing, bullets and shot-gun pellets, alloyed with tin to make pewter drinking vessels, and as a weight. From the 1920s it was used as an additive to petrol (gasoline) to improve the performance of the combustion engine, but this practice has almost ceased because of pollution concerns.

The main uses of lead at present, apart from roofing, are in lead-acid electric storage batteries, and as a shield against X-rays. Many former uses have had to be abandoned because of the element's toxicity. It is poisonous to all animals, damaging the nervous system in particular, which has caused its use to be restricted by law in many countries. Its use in lead water pipes is thought to have been particularly harmful in the past, but most pipes have now been replaced by copper or plastic ones.

The hardness of lead can be improved by adding small proportions of copper, tin or other elements.

Occurrence and origin

Lead ores occur as both sulphides (**galena** – Fig. 5.3B) and carbonates (**cerussite**) and are typically found along with zinc, copper and silver.

The crustal abundance of lead is only about 14ppm; it makes up less than 1ppm in oceanic basalts, assumed to be the original source, and occurs in very low quantities in most sedimentary and igneous rocks. However, its concentration in the average worked ore body is relatively high, between 4% and 10%.

Figure 5.3B Crystals of galena (lead sulphide)
Image credit: © Shutterstock, BrankoG.

By far the most common source of lead ore is limestone (or dolomite) strata, where it occurs in veins or cavities associated with calcite or **barite** (barium carbonate) in Mississippi Valley-type deposits or in SEDEX-type ore bodies within black shales, such as Mount Isa in Australia. The ore bodies are often concentrated alongside limestone reefs or other lithological boundaries marking changes in permeability (see Fig. 4.2A). Most ore bodies are not directly associated with igneous activity, and the processes involved in their concentration are considered to be sedimentary deposition followed by aqueous transport via brine solutions, which are capable of dissolving the lead from large volumes of sedimentary rock.

Many deposits seem to have been emplaced during periods of orogeny, especially during the late Carboniferous, when the ore fluids may have been driven out of the source rocks by tectonic activity.

Resources

World production in 2012 was 5.5 million tonnes, about half of which came from recycled scrap. The top five producing countries are: China, Australia, USA, Peru and Mexico (table 5.2B). According to one estimate, at the current rate of use, reserves will last for 42 years. However, concerns about toxicity and environmental pollution have significantly reduced the rate of consumption, and this process will undoubtedly continue, particularly if an alternative to the lead storage battery is widely adopted.

Zinc

Zinc (Zn) (30) is an essential trace element for both animals and plants, and is a constituent of many important enzymes that control development, the immune system and fertility. Adequate levels of zinc in the soil are essential for all the main cereal crops – rice, wheat and maize. It is hard and brittle at low temperatures and its main uses are as alloys and in compounds.

History of use

Brass, an alloy of copper and zinc, has been made since at least 1000BCE and was widely used by the Romans. Metallic zinc was produced in India in the twelfth century but not isolated in the West until the late seventeenth century. The galvanic cell, pioneered by Luigi Galvani and Alessandro Volta, in which copper and zinc plates are separated by an electrolyte, is a method of producing, or storing, electricity, and is the basis for much of the modern zinc production.

Over half of the zinc produced at present is used as an anti-corrosion coating on steel, i.e. in **galvanized** steel. Zinc is also widely employed in the manufacture of a large number of die-cast products, alloyed with copper, aluminium or magnesium; because of their low working temperature, these can be easily cast into intricate shapes. Zinc compounds such as zinc oxide have many other industrial applications. Zinc supplements are used to counteract zinc deficiency in the diet, and zinc oxide preparations in ointments to protect the skin.

Occurrence and origin

Zinc is relatively abundant in the crust, at between 70 and 80ppm. Zinc ore occurs mainly as the sulphide, **sphalerite**, which is almost always found in association with sulphides of lead, copper and iron, or the carbonate, **smithsonite**. Most zinc ore is sourced from carbonate-hosted lead-zinc deposits (see Fig. 4.2A).

The zinc production process produces large quantities of contaminants, including sulphur dioxide vapour and heavy metals. Also, high levels of zinc in soil or in rivers flowing through mining areas are dangerous to wildlife.

Resources

Zinc is the fourth most commonly used metal (after iron, aluminium and copper). World production in 2012 was about 12 million tonnes, the five leading producing countries being China, Peru, Australia, India and the USA (table 5.2C). Identified reserves then totalled 250 million tonnes and are estimated to last, at present rates of exploitation, until between 2027 and 2055. The largest reserves are in Iran, Australia, Canada and the USA.

Tin

Tin (Sn) (50) is a soft, ductile, silvery-white metal, one of the earliest metals to be used by humans, after it was discovered that by alloying it with copper, the much stronger bronze was produced, heralding the Bronze Age, at around 3000BCE.

History of use

Tin was mined in several countries around the Mediterranean, especially in Britain, and was widely traded during Classical Greek and Roman times. During the medieval period, it was employed as tin plating on iron and also alloyed with copper, lead and antimony in pewter, widely used in drinking vessels. In more modern times, one of its main applications in addition to tin plating is as **solder** for joining pipes, and in electrical circuitry. The deaths of Captain Scott and his Antarctic exploration team have been attributed in part to the low-temperature decay of the tin solder that sealed their fuel containers, allowing the fuel to leak out.

Occurrence and origin

Tin makes up about 2.3ppm in the crust and occurs mainly in the form of the black ore **cassiterite** (tin oxide), although it also occurs in the form of sulphides. The tin originates as a late-stage hydrothermal product of granitic magmas (see Fig. 4.1A), from mineral veins around the granite plutons, as in the historic Cornish deposits. Similar deposits have been worked in SE Asia, Nigeria and Bolivia, but the metal is largely sourced now from alluvial (placer) deposits.

Resources

World production in 2012 totalled 240 kilotonnes, and estimated reserves were 4.7 million tonnes. The top five producing countries are China, Indonesia, Peru, Bolivia and the Democratic Republic of Congo (table 5.2D).

Tungsten

Tungsten (W) (74) is one of the heaviest metals, and is the strongest at high temperatures, with a melting point of 3422°C, which explains its popularity in the armaments industry. The name tungsten is derived from the Swedish 'heavy stone', which was applied to the calcium tungstate ore subsequently called **scheelite**. The symbol W comes from the name of the other common tungsten ore, **wolframite**.

History of use

Tungsten was first isolated and named by the Spanish brothers Juan José and Fausto Elhuyar in 1783. The ductility of this metal allows it to be drawn out into very fine wire, which led to the introduction of the tungsten filament light bulb in 1906; these have now been largely replaced by more efficient types – in the tungsten bulbs only 10% of the energy was converted into light.

Tungsten played an important role in both the first and second world wars, for example in high-velocity anti-tank shells. During the First World War, Germany was able to effectively deploy its enormous howitzer guns, which could fire shells for distances of over 12km, only because the use of tungsten steel enabled them to keep firing without overheating. The supply of tungsten was critical during the Second World War, when the major source was Portugal which, to demonstrate its neutrality, supplied vast quantities to both sides in the conflict!

Tungsten is mainly employed now in steel alloys such as tungsten carbide, which is strong enough to cut cast iron, and is used for cutting and drilling tools.

Occurrence and origin

Crustal abundance of tungsten is about 1–1.25ppm. The principal ores are scheelite (calcium-tungsten oxide) and wolframite (iron-manganese-tungsten oxide); these are found mainly in the late-stage volatile-rich fractions of granitic magmas, often associated with tin ore. The recently-opened Drakelands Mine in Devon, SW England, is designed to exploit the largest known tungsten deposit in Europe.

Resources

World production of tungsten in 2012 was estimated at 75.7 kilotonnes, mostly from China. Significant quantities were also produced by Russia, Canada and Bolivia (table 5.2E). Data from the USA are not available. Tungsten is regarded as one of the key 'strategic' minerals. 2012 reserves were estimated at 3,100 kilotonnes, but world resources have been estimated at around 7 million tonnes; between 35% and 40% is currently being recycled.

Titanium

Titanium (Ti) (22) is the ninth most abundant element in the Earth and one of the most useful, being as strong as steel and 45% lighter.

History of use

Titanium was discovered in 1791 by William Gregor and named a few years later by Martin Klaproth after the God 'Titan'. However, it has only been widely exploited in recent times. Because of its strength, lightness, and corrosion resistance, and since it is unaffected by metal fatigue, it is much used in the aviation industry for aircraft mainframes and engines, and in the manufacture of sports goods such as tennis rackets and bicycles. Its corrosion resistance has led to many maritime applications as well. It is also non-toxic, and is ideal for replacement body parts.

Occurrence and origin

The crustal abundance of titanium is c.0.56–0.66%, and its oxide is found in most igneous and metamorphic rocks; basic igneous rocks contain around 2%. The two principal ores are **rutile** (titanium oxide) and **ilmenite** (iron-titanium oxide). Mined deposits of these ores are mainly sourced from heavy mineral sands such as old beach deposits found, for example, in coastal regions of Australia.

Resources

World production in 2012 totalled 200 kilotonnes. The top five producing countries were China, Russia, Japan, Kazakhstan and the Ukraine (table 5.2F). Total reserves were estimated to be about 600 million tonnes.

Zirconium

Zirconium (Zr) (40) is a light, soft, silver-grey metal, whose silicate, **zircon**, forms finely-shaped crystals; its transparent varieties are employed as a substitute for diamond, from which it is hard to distinguish it, except by the expert!

History of use

Zirconium was discovered and named in 1789 by Martin Klaproth, but not finally isolated until 1824. Apart from the use of zircon in jewellery, there are many industrial applications of zirconium due to its light weight, heat tolerance and corrosion resistance. For these reasons, it has been employed in its oxide form as a ceramic material in the manufacture of jet engines and space vehicles, and because it is unaffected by neutron bombardment, as cladding in nuclear reactors.

Occurrence and origin

The crustal abundance of zirconium is estimated at 130–250ppm; it occurs as silicates, especially zircon, in oceanic basalts in proportions of 50–300ppm, but is present in larger concentrations in most acid igneous rocks and, particularly, in **pegmatites**. It is sourced in the form of zircon (zirconium oxide) as a by-product of the mining of the placer deposits of the titanium minerals ilmenite and rutile, and to a lesser extent, tin (cassiterite).

These ores occur as heavy mineral concentrates, mainly in beach sands.

Resources

World production of zirconium in 2012 totalled about 1460 kilotonnes, the top producing countries being Australia, South Africa, China and Indonesia (table 5.2G). World reserves are considerable – estimated at 67 million tonnes. Much of the metal can be easily recycled.

Manganese

Manganese (Mn) (25) is a silvery-grey metal similar in appearance to iron. The element is hard and brittle, but when alloyed with iron makes a particularly hard steel with many applications, for example, in railway tracks and safes.

History of use

Black manganese oxide has been known since prehistoric times, when it was used as a pigment in cave paintings, and both the Egyptians and the Romans used it as a colouring agent in glass-making. It was first isolated in 1774 by Johan Gahn. At present, between 85% and 90% of the manganese produced is employed in steelmaking, but it is also alloyed with aluminium to improve corrosion resistance, and manganese oxide is widely used as a cathode in zinc-carbon and alkaline batteries.

Occurrence and origin

Manganese is the twelfth most abundant element, making up around 0.1% of the crust, and up to 0.2% of the average ocean-ridge basalt. It occurs widely on the ocean floor in the form of manganese nodules. The principal

Figure 5.4 Rhodochrosite (manganese carbonate)
Image credit: © Shutterstock, PNSJ88.

ore is **pyrolusite** (manganese dioxide) which occurs widely; it also occurs as the beautiful pink manganese carbonate **rhodochrosite** –(Fig. 5.4) and as sulphates, chlorides and silicates.

Manganese ore, principally oxides and carbonates, occurs in a wide variety of geological environments. Although primarily of volcanic origin, most deposits are not directly linked to igneous bodies, but result from secondary processes of enrichment, either via hydrothermal activity leading to veins and breccia infilling, or by sedimentary deposition. Most worked ores are from sedimentary layers, such as the extensive Ukrainian deposits, or residual soils.

Resources

World production in 2012 totalled 15,800 kilotonnes. The top five producing countries were Australia, South Africa, China, Gabon and India (table 5.2H). Around 80% of known resources are in South Africa, but Ukraine, Australia, India, China, Gabon and Brazil also host large deposits. The general abundance of manganese means that, although there are limits to economically viable reserves, the potential resources are theoretically unlimited.

Chromium

Chromium (Cr) (24) is a hard, lustrous, silvery metal, the twenty-first most abundant element in the crust. It is most familiar for the shiny chrome-plating on the car bumpers of the 1950s and '60s.

History of use

Chromium oxide was exploited by the Chinese in the third century BCE as a coating on the metal weapons of the warriors of the Terracotta Army, and chromium compounds have been known as pigments for many centuries. However, the metal was not isolated until 1797, when Nicolas Vauquelin obtained it from the bright red ore **crocoite** (lead chromate). Its main application now is in the manufacture of stainless steel and other steel alloys, which may contain up to 25% chromium, and in chrome-plating. Some highly coloured chromium compounds (e.g. chrome yellow) are still employed as pigments, although their use has declined due to environmental concerns. Trace amounts of chromium are essential for animal metabolism, but some chromium compounds are highly toxic – chromate dust is carcinogenic, and there is a continuing pollution problem from chromate residues in former industrial sites.

Occurrence and origin

The crustal abundance of chromium is estimated at 100–350ppm, and the element is present in about the same proportions in oceanic basalts. Chromium occurs naturally, mainly in the form of oxides and in combination with other metals; small amounts of chromium added to corundum give the red colour to rubies (Fig. 5.2A) and the green to emeralds.

The main industrial source of chromium is from deposits of **chromite** (iron-chromium oxide) which occur as layers in ultrabasic intrusions such as the Bushveld Complex in South Africa,

as pods in mantle-derived peridotites, or as a sedimentary deposit in heavy mineral sands.

Resources

World production in 2012 totalled 25.6 million tonnes, the main producing countries being South Africa, India and Kazakhstan (table 5.2I). There are still considerable resources in these countries, as well as in Russia and Turkey. In 2012, world reserves were estimated at 480 million tonnes.

Vanadium

Vanadium (V) (23) is a hard but ductile, shiny, silver-coloured metal, whose main application is in the production of a very strong and light type of steel with good corrosion resistance.

History of use

Vanadium was discovered (for the second time!) and named in 1831 by the Swedish chemist Nils Sefström. It has been employed by the car industry since 1913 because, when alloyed with iron, it makes a much stronger and lighter steel. It has many other industrial uses: for example, in the manufacture of protective armour and high-speed drill tools. It is also alloyed with aluminium and titanium in jet engines. It is abundant in soils and absorbed by most plants, and is an essential trace element for the human body; however, vanadium pentoxide dust is considered to be hazardous to the respiratory system. Certain marine organisms concentrate vanadium from sea water, which may explain the large concentrations of the element in fossil fuel deposits.

Occurrence and origin

The crustal abundance of vanadium is about 120–190ppm, and it is present in oceanic basalt in proportions of between 100 and 300ppm. Vanadium occurs naturally in many different compounds, the most important economically being **patronite** (vanadium sulphide) and various compounds containing vanadium pentoxide (e.g. **vanadinite**). Much commercial vanadium is sourced from concentrations of vanadium-bearing magnetite in ultrabasic intrusions.

Resources

World production in 2012 totalled 74 kilotonnes. Almost all comes from three countries: China, South Africa and Russia (table 5.2J). It can also be obtained from flue dust derived from the burning of fossil fuels, and as a by-product of uranium mining of bauxite ore. Estimated reserves were around 14 million tonnes.

Tantalum

Tantalum (Ta) (73) is one of the heavier metals, with a silvery-grey shiny appearance. It was named after the mythological King Tantalus, who was banished by the gods and condemned to be tormented for eternity by being forced to stand up to his neck in water, which he was unable to drink! The name is apt, since it proved equally frustrating to isolate the element because of its similarity to niobium – the two were finally separated in 1842 by Heinrich Rose.

Uses

Tantalum is highly resistant to corrosion because of a thin surface layer of oxide, which also serves as an insulator: hence its use in tiny capacitors for electrical storage in devices such as mobile phones and radios. It has also found an application in surgical procedures because, like niobium, it does not react with human tissue.

Occurrence and origin

The crustal abundance of tantalum is around 2ppm. Its main sources are the minerals **tantalite** and **coltan**. Tantalite is an iron-manganese-tantalum oxide, whereas coltan is a mixture of tantalite and **columbite** (see niobium, below). Tantalite ores are found in highly fractionated acid pegmatites and in **carbonatites**. Most tantalite is mined in Brazil, Mozambique and Rwanda. However, illegal mining of coltan in the Democratic Republic of Congo is notorious because of its role in funding insurrection, and also because of the associated destruction of the habitat of the endangered Mountain Gorilla. Nevertheless, some coltan production undoubtedly finds its way into world markets via neighbouring Central African countries.

Resources

World production of tantalum in 2012 totalled 760 tonnes, mainly from Brazil, Mozambique and Rwanda (table 5.3A) but has recently varied considerably from year to year. Global reserves have been estimated at over 100 million tonnes, mostly in Brazil and Australia, and there is no significant problem with resources. About 20% of the tantalum used is currently recycled.

Niobium

Niobium (Nb) (41) is a soft, shiny, grey metal; it was named after Niobe,

Greek goddess of grief, and daughter of Tantalus, because it was discovered from an ore of tantalum.

History of use

It was discovered in 1801 by Charles Hatchett, but it was not until 1846 that Heinrich Rose gave it the name niobium, and it was finally isolated in 1864 by C.S. Blomstrand. It is now used in high-grade steel to give improved strength and malleability, and in combination with iron, cobalt and nickel to make super-alloys for gas turbines and

rocket components. Another important application is in super-conducting materials when alloyed with nickel-germanium or nickel-tin. It has no known biological role and does not react with human tissue, making it ideal for pacemakers and surgical implements.

Occurrence and origin

The crustal abundance of niobium is about 20ppm. It occurs in alkaline igneous rocks and is recoverable from highly fractionated acid pegmatites as the minerals

country	amount in kilotonnes
Ruanda	150
Brazil	140
Congo	100
Ethiopia	95

A. Tantalum

country	amount in kilotonnes
China	7.3
Korea	3.0
Mexico	1.624
Canada	1.1
Kazakhstan	1.3

D. Cadmium

country	amount in tonnes
China	405
R. of Korea	165

G. Indium

country	amount in kilotonnes
Brazil	45
Canada	4.7

B. Niobium

country	amount in kilotonnes
China	1,350
Kyrgyzstan	250
Chile	52
Russia	50
Peru	40

E. Mercury

country	amount in tonnes
Australia	27
South Africa	21
China	7

H. Hafnium

country	amount in kilotonnes
China	7,000
Mexico	940
Canada	121

C. Bismuth

country	amount in tonnes
China	141
Germany	35
Kazakhstan	25

F. Gallium

country	amount in tonnes
Chile	27
USA	7.9
Poland	6
Uzbekistan	5.4

I. Rhenium

Table 5.3 Metal production by main producing countries – III
Data sources: US Geological Survey (USGS): A-E, G, I, 2014; F, H, 2012.

columbite (iron-manganese-tantalum-niobium oxide) or **pyrochlore** – a complex hydroxide of tantalum and niobium, along with sodium, calcium and fluorine.

Resources
World production of niobium in 2012 totalled around 50.1 kilotonnes, mainly from Brazil and Canada (table 5.3B). World reserves are plentiful, estimated at around 4.3 million tonnes.

Bismuth
Bismuth (Bi) (83) is a heavy, silvery, post-transition metal which sits between lead and polonium in the Periodic Table but, unlike them, is non-toxic and has had a wide variety of functions since at least the fifteenth century, including its application as a medicine for stomach ailments.

History of use
During the fifteenth century, bismuth was exploited by alchemists, although it was not recognized as a separate metal until 1753, when it was distinguished from lead and tin by Claude Geoffroy. Alloyed with lead, it was used in hot-metal typesetting because of its low melting temperature. Since then it has found a wide variety of applications: as a medicine, in cosmetics, as a replacement for lead in hunting ammunition, and to produce the sparkle in fireworks! Bismuth vanadate (a bismuth-vanadium oxide) is a yellow pigment which has replaced the poisonous lead chromate.

Occurrence and origin
The crustal abundance of bismuth is very low, less than 0.03ppm. It occurs both as the native element and as the sulphide, **bismuthinite**, the oxide, **bismite** and the carbonate, **bismutite**. Bismuth occurs in veins associated with the ores of tin, silver, copper and nickel in IOCG-type deposits (see Figure 4.1A). Most bismuth comes from the sulphide ore as a by-product of copper and tin smelting.

Resources
World production of bismuth in 2012 was estimated at 8.2 kilotonnes, the top three producers being China, Peru and Mexico (table 5.3C). Reserves were estimated at around 320 kilotonnes.

Cadmium
Cadmium (Cd) (48) is a soft, silvery metal that can be easily cut with a knife. It is so toxic that it is listed by the UN Environmental Programme among the top ten hazardous pollutants. It achieved notoriety in Japan in the mid-twentieth century when cadmium pollution from a zinc mine caused rice grown in the area irrigated by polluted water from the mine to acquire ten times the normal levels of cadmium. This led to the widespread occurrence of a disease known as itai-itai ('ouch-ouch') which caused severe joint pain and was finally recognized in 1946 as a cadmium-related disease.

History of use
Cadmium occurs together with zinc in zinc ore, and the two metals are closely related. Cadmium was discovered in the early 1800s by German chemists attempting to obtain zinc from smithsonite (zinc carbonate), and was eventually separated independently by Friedrich Stromeyer and K.S.L. Hermann.

Cadmium sulphide, known as 'cadmium yellow', was formerly much used as a pigment and more recently in the nickel-cadmium and alkali-cadmium batteries and for electroplating, but these have largely been replaced by more benign alternatives.

The high toxicity of cadmium is due to the fact that it can readily replace zinc, which is an essential mineral in the human body; however, unlike zinc, the cadmium is stored and can become concentrated to dangerous levels. Since many of the former uses of cadmium have been banned or discontinued, most exposure to cadmium now comes from tobacco smoking.

Occurrence and origin
Crustal abundance of cadmium is very low, around 0.15ppm. It occurs combined with zinc in its sulphide, carbonate and silicate ores, and is produced as a by-product of zinc mining.

Resources
World production in 2012 was estimated at about 20 kilotonnes, the top six producers being China, Korea, Japan, Mexico, Kazakhstan and Canada (table 5.3D). World reserves were estimated at c.500 kilotonnes. Considerable amounts of cadmium are recycled from cadmium batteries and other sources.

Environmental concerns
The Japanese experience has led to strict regulations governing the use of cadmium, which is a restricted chemical under the EU REACH regulations (restriction, evaluation, authorization and restriction of chemicals).

Mercury

Mercury (Hg) (80) is the only metal that is liquid at ordinary temperatures and pressures, and being one of the heaviest metals, is so dense that even lead will float on it. It is familiar as the silvery liquid of the thermometer and barometer – used for these instruments because of its rapid response to small changes in temperature and pressure, and because of its low freezing point (-39°C) and high boiling point (357°C).

History of use

Mercury was known by the Romans, who called it *hydrargyrum* (from the Greek for 'water-silver'), hence the symbol Hg; it is also popularly known as 'quicksilver'. The oldest known sample of mercury was found in an Egyptian tomb dated at 1600BCE, but the red mercury ore **cinnabar** was used in prehistoric times in cave paintings. Another ore, **calomel** (mercury chloride) was extensively employed to treat various conditions, including venereal disease, until, in the twentieth century, it was discovered to be highly toxic. It is believed that both Henry VIII of England and Ivan the Terrible of Russia were treated by it. Prolonged exposure to mercury leads to damage to the brain and central nervous system, and its former use by hat-makers gave rise to the expression 'mad as a hatter'!

It was first identified as a metal in the mid-eighteenth century by the Russian scientists A. Braun and M.V. Lomonosov, who were surprised to find that it solidified in their thermometers in extreme cold. Although its use in thermometers has largely been superseded, it continues to find a role in mercury vapour lamps and in various other niche applications.

The Spanish used large amounts of mercury to extract gold and silver from their mines in South and Central America in the sixteenth and seventeenth centuries because of the ease with which mercury formed alloys with the precious metals, enabling them to be extracted from the crushed host rock. Illegal gold-panning with mercury remains a problem in some countries. Due to the high toxicity of mercury, many historical applications of the metal have been abandoned, and are now illegal in many countries including the USA and the EU; it has been banned completely in both Norway and Sweden.

Occurrence and origin

Crustal abundance of mercury is extremely low, below 0.1%, but it can be enriched to as much as 14% in the commonest mercury ore, cinnabar (mercury sulphide). Most cinnabar is found in quartz veins associated with recent or active volcanic areas, for example in hot-spring deposits. Historically, the Almeda mine in Spain produced the largest amounts of mercury, but today, China is the top producer. Many old mines in America and Europe have either been exhausted or have been shut down because of environmental concerns.

Resources

World production of mercury in 2012 totalled 1,810 tonnes, the top three producers being China, Kyrgyzstan and Peru (table 5.3E). The use of mercury has halved in the last decade and there are large stocks. About 30% of present production consists of recycled mercury, and this proportion will undoubtedly increase because of the restrictions in its use.

Environmental concerns

The high toxicity of mercury has led the World Health Organization to designate mercury as an 'occupational hazard' with specified exposure limits. Mercury occurs in the atmosphere in measurable quantities from volcanic eruptions, but there have been significant increases due to pollution from coal-fired power stations and other sources that can be a health hazard. Leakage of mercury compounds from old mines and waste dumps causes serious concern and is strictly monitored in many countries. The dumping of mercury in Minamata Bay in Japan, which led to its concentration in fish and shellfish, caused widespread poisoning of the local population. This incident led to the introduction of the **Minamata Convention** on Mercury by the UN Environmental Programme, signed in 2013, which specifies safe limits and practices in dealing with mercury and bans the opening of new mines.

The rare metals

The following group of metals occur naturally in very small quantities, are difficult to extract, and have few industrial applications – though some of these are quite important. The amounts produced are therefore minute (measured in tonnes or kilograms rather than kilotonnes) compared to the majority of the metals previously discussed. They are similar in many respects to the rare-earth elements (lanthanides) described in chapter 7.

Scandium (Sc) (21) is the first of the transition-metals group of the Periodic

Table. Although not particularly rare in the crust (*c.*22ppm.), it only occurs in small quantities in a few minerals and is produced as a by-product of uranium and rare-earth mining. It was first isolated in 1879, by Lars Nilson. The only important application of scandium is in the aircraft industry, where it can improve the strength of aluminium by the addition of small quantities of around 0.1%. Global production is probably less than 2 tonnes annually, from China, Russia and Ukraine. There are considerable stockpiles and available resources are theoretically large.

Gallium (Ga) (31) is unusual for a metal in being so soft that it will melt in the hand, and remains liquid up to 2373°C, making it ideal for high-temperature thermometers. It has featured in an amusing conjuring trick: an unsuspecting guest is given a gallium spoon to stir his tea, upon which the spoon immediately dissolves!

Gallium was discovered in 1875 by Paul de Boubaudron by extracting it from the aluminium ore bauxite, which contains it in small quantities (*c.*50ppm). Crustal abundance is about 19ppm. It is now produced as a by-product of zinc or copper refining. It is used in semiconductors for supercomputers and mobile phones, and its radioactive isotope is used in cancer treatment.

Total world production in 2012 was estimated at 383 tonnes, the top three producers being China, Germany and Kazakhstan (table 5.3F). There are large potential resources in bauxite ore, which are uneconomic to extract now, and much of the metal used can be recycled.

Yttrium (Y) (39). An ore containing yttrium was originally found near the village of Ytterby near Stockholm, after which the element was named, and it was isolated in 1828 by Friedrich Wöhler. Yttrium is very similar in its properties to the rare earth minerals and is found in the same ore deposits. Although its crustal abundance is as much as 33ppm, it occurs in very few minerals – the most important being the mineral **xenotime** (yttrium phosphate) which contains 60% of yttrium. Yttrium is employed to make synthetic garnets for lasers, and in superconductors. A radioactive isotope of yttrium is used in cancer treatment. World production in 2012 was 8900 tonnes, of which China was the main source. World resources are believed to be large.

Indium (In) (49) is a soft, lustrous, silvery metal, discovered accidentally in 1863 by Ferdinand Reich and Hieronymus Richter when analysing the zinc ore sphalerite. Although it does occur in the native form, most production is obtained as a by-product of zinc smelting. It was first used industrially in the Second World War to coat bearings in high-performance aircraft engines; however, the principal application now is in thin-screen television and computer monitors because of its importance as a transparent conductor of electricity. A radioactive isotope is used medically in scanning devices.

Crustal abundance is estimated at about 0.25ppm and total world production in 2012 was estimated at 782 tonnes, mostly from China and the Republic of Korea (table 5.3G).

Hafnium (Hf) (72) was one of the last of the stable (i.e. non-radioactive) elements to be discovered because of its similarity to zirconium, occurring with it in the mineral zircon in proportions of 1–4%. It was finally isolated in 1923 by the Danish scientists Dirk Coster and Georg von Hevesy using X-ray techniques. It is sourced along with zirconium, mainly from heavy mineral sands.

One of the most important applications of hafnium is in the manufacture of the control rods of nuclear reactors, because of its ability to absorb neutrons. It is also used in micro-processors, and in high-temperature alloys for rocket engines and cutting tools.

The estimated crustal abundance of hafnium is about 3ppm but it is difficult and expensive to extract, and only 64 tonnes are produced annually, mainly by Australia, South Africa and China (table 5.3H).

Rhenium (Re) (75) has one of the highest melting points of all the elements (only tungsten and carbon have higher) and was the last stable element to be discovered. It was finally isolated only in 1925 by Walter Noddack and Ida Tacke from molybdenite, which contains tiny quantities (*c.*0.0015%) of the metal. It is one of the rarest elements (the estimated crustal abundance is 0.0007ppm!) and does not occur as a free metal, so is consequently very expensive to produce. It is used in nickel-iron alloys to strengthen turbine blades in military fighter-jet engines, which accounts for most of the global production; it is also used in thermocouples, and as a catalyst in hydrogenation reactions involved in the petrochemical industry.

Total world production of rhenium in 2012 was 52.6 tonnes, mostly from Chile, as a by-product of copper-molybdenum mining, but also from the USA, Peru and Poland (table 5.3I). There are potentially large resources, the main problem being the costs of recovery. Much can be recycled from discarded turbine blades.

Thallium (Tl) (81) has achieved notoriety as an 'undetectable' poison because its symptoms (vomiting, delirium and hair loss) can be confused with other causes. The poisonous substance is thallium sulphate, which dissolves in water, and is colourless, tasteless and odourless. Widely used in the past as a rat poison, it is now banned in most countries. In a famous case of thallium poisoning, Australian grandmother Caroline Grills was convicted in 1953 of killing four members of her family before being detected by a suspicious relative. Her fellow prison inmates called her 'Aunt Thally'!

More orthodox uses of the metal are as superconductors in the electronics industry and in pharmaceuticals.

A radioactive isotope is employed medically in nuclear cardiography.

Thallium was discovered independently in 1861 by William Crookes and Claude-Auguste Lamy. The metal occurs mostly in the minerals **sylvite** (potassium chloride) and **pollucite** (a complex alumino-silicate); trace amounts are present in copper, zinc and lead ores. Crustal abundance is estimated at $c.0.5–0.8$ppm, and world resources at 17 tonnes. Total world production in 2012 was estimated at 10 tonnes, much of it from Kazakhstan and China.

Technetium (Tc) (43) was the first radioactive element to be created artificially; it was produced in 1937 by Emilio Segrè and Carlo Perrier by bombarding a sample of molybdenum with deuteron particles in a cyclotron. Trace amounts are found in uranium ores but most technetium is produced in nuclear reactors. The most stable isotope has a half-life of only four million years, so that none of the metal originally present in the crust has survived.

Technetium is used in medical imaging to study brain, lung and bone tissue. If combined with tin, it binds with red blood cells to enable circulatory problems to be traced. Samples for medical use are produced from spent fuel rods from nuclear reactors. Most of the world supply of technetium comes from reactors in Canada and the Netherlands.

Although its toxicity is low, inhalation of technetium dust particles is hazardous, and because of its long half-life, waste disposal is potentially problematic.

Metallic mineral resources – II

The precious metals

Gold and silver have been regarded as particularly special metals since the very earliest times, when they were used for ornamental purposes because of their great beauty, resistance to wear or decay, and relative ease of extraction from the host rock. More importantly, they have been employed both as currency and as a store of value. They are therefore often termed the 'precious metals'. To this category are often now added two of the platinum-group metals – platinum and palladium.

Gold

Gold (Au) (79) is unique among the elements as a symbol of beauty, power and wealth, and of the highest aspirations and achievements, hence the award of gold medals in the Olympic Games and the Nobel Prize. It was esteemed by the earliest civilizations – one of the oldest gold artefacts, found at a site in Bulgaria, was dated at *c.*4700BCE. It was also prized by the ancient Egyptians – Figure 6.1 illustrates the gold death mask of King Tutankhamun. The symbol Au comes from the Latin *aureum* for gold.

Currency

The first gold coins known were produced in Lydia (in modern Turkey) around 600BCE, and gold currency was widely circulated in Greek and Roman times. In Western European countries, gold replaced silver as coinage during the thirteenth and fourteenth centuries, and when paper money began to circulate as legal tender by governments in the nineteenth century, it had to be backed by an equivalent amount of gold, which was stored as a gold reserve in the form of **bullion** (e.g. gold bars). Countries that adopted this system were said to be on the 'gold standard'. Rigid adherence to the gold standard proved to be too restrictive to the major powers during times of economic pressure, such as the periods of the first and second world wars, and the gold standard was finally abandoned in 1975. However, many countries still keep a store of gold – the US Central Bank is reputed to hold 3% of all the gold ever mined!

Gold is regarded as a store of wealth and a hedge against decreases in the value of the currency, and is widely traded in the commodities market. Despite this, its value against the major currencies fluctuates widely. It is bought and sold in the form of gold bars or gold coins such as the **kruggerrand**. In 2014, a kruggerrand would set you back £821 and a British gold sovereign (which originally had a face value of £1) would now cost £390!

Other uses

Gold is one of the least reactive metals, being unaffected by exposure to air, water, alkalis and most acids, which explains its value in ornamentation and dentistry, for example. It is also one of the most ductile metals and can be beaten into very thin (even transparent) sheets, or drawn into fine wire or thread. High electrical conductivity has led to its application in electrical connectors and as a coating on plastics and glass. Much of the gold used in ornamentation is in the form of alloys: the addition of copper gives 'red' gold, and silver, 'white' gold, for example. About 50% of the gold produced is destined for jewellery, much of it for India and China; another 40% is traded as investment and hoarded; and the remaining 10% is used for various industrial purposes, such as electronics and medicine.

Occurrence and origin

Gold is extremely rare in the crust, around 0.004ppm. It typically occurs in the native state in quartz veins, usually alloyed with silver, along with iron and copper sulphides, in IOCG deposits (see Fig. 4.1A, B) and, more conveniently, in placer deposits derived from the erosion of gold-bearing veins. The lust for gold has led to many famous 'gold rushes', such as those of California, Australia, the Klondike in Canada, and the Witwatersrand in South Africa, all in

Figure 6.1 Gold death mask of King Tutankhamun
Image credit: © Shutterstock, mountainpix.

the mid-to-late nineteenth century. Even today, about a quarter of mined gold comes from small-scale mining, much of it illegal. Ore grades as low as 0.5ppm can be mined profitably in surface deposits but grades of under 3ppm would usually be unprofitable in deep mines. Gold can be directly hand-picked from placer deposits by '**panning**', however in commercial gold mining, the ore is crushed and extracted by dissolving it in potassium or sodium cyanide.

Resources

Total world production of gold in 2012 was around 2700 tonnes, the top four producers being China, Australia, the USA and Russia (table 6.1A). Gold can be indefinitely recycled with negligible loss, and there is probably enough gold in circulation to satisfy any likely industrial demand, so that demand for new gold for investment or ornamental purposes in the future will be controlled by the price of extraction, and is not therefore a serious resource concern.

Environmental concerns

Although modern mining is strictly controlled, there are severe pollution problems with many abandoned mines, and also with contemporary unofficial and unregulated mining because of the sodium cyanide or mercury used for processing. Also, old mine dumps often contain dangerous heavy metals such as cadmium, lead and arsenic.

Silver

Silver (Ag) (47) is a soft, white, shiny metal, familiar because of its use in jewellery, commemorative trophies and coins. It is slightly reactive to sulphur in the air, which produces a greyish tarnish. The symbol Ag is derived from the Latin *argentum*. Silver has been exploited since at least 3000BCE for decorative purposes.

Currency

Coins made of silver-gold alloy were first used by the Lydians, and silver coins were the standard currency throughout the Roman Empire. The Spanish conquest of South America in the sixteenth century led to enormous quantities of silver being released onto the European market, causing a collapse in the value of silver and an effective devaluation of the silver currency, which was eventually replaced by gold by the beginning of the nineteenth century.

Other uses

Silver has been widely employed for ornamental purposes, usually alloyed with copper: **sterling silver** in trophies contains 93% silver, tableware contains 80%. It is also used with nickel as a coating in electroplating – thus, 'electroplated nickel silver' (epns). Other destinations include reflective backing for mirrors, silver iodide for seeding clouds to precipitate rain, and silver oxide for watch batteries.

Occurrence and origin

Crustal abundance of silver is very low (*c*.0.08ppm) and although native silver was mined originally, the element is now found typically in the form of sulphides and chlorides: **argentite** (silver sulphide), **pyrargentite** (silver-antimony sulphide) and **chlorargyrite** (silver chloride), along with lead, copper and arsenic ores in porphyry copper, IOCG and SEDEX deposits (see Fig. 4.1A). Silver is now produced as a by-product of the electro-refining of other metals such as copper, nickel, lead and zinc from such deposits.

Resources

World production of silver in 2012 totalled 25.5 kilotonnes, the top three producers being Mexico, Peru and China (table 6.1B). Reserves were estimated by the US Geological Survey in 2012 at 520 kilotonnes, which at present rates of extraction would only last about 20 years. However, essential uses of silver are quite limited, and much of it can be recycled.

The Platinum Group metals

Platinum (Pt) (78) is one of the rarest elements – considered to be more desirable even than gold – hence the 'platinum' credit card has a higher status than the gold one. It is one of a group of four rare elements that occupy the same space in the Periodic Table, share very similar properties, and are often found together: platinum, palladium, osmium and iridium. Platinum is highly resistant to wear or tarnish and is valued in jewellery for that reason. It has also found several niche industrial applications: e.g. the catalytic converters in automobiles, and as the basis of the chemotherapy drug cisplatin. The international standard metre and kilogram are made from platinum because of its stability.

Platinum was known to the ancient Egyptians and the Romans: an early example, dated at the seventh century BCE and attributed to the Egyptian Queen Shapenapit, was found at Thebes. The name comes from the Spanish platina (little silver) following their discovery of the metal in Columbia.

The crustal abundance of platinum is extremely low – around 0.005%, but it is much commoner in meteorites. It occurs both as the native metal and alloyed with iridium, and in nickel deposits as a sulphide or arsenide (**sperrylite**). It is mined in some of the major magmatic nickel-copper-platinum group deposits (see Fig. 4.1C) such as Bushveldt, Sudbury and Noril'sk, where it is obtained as a by-product of the electro-refining of copper and nickel. World production of platinum in 2012 was 183 kilotonnes, over 70% of it from South Africa (table 6.1C). Reserves of platinum group metals were estimated at 66 million tonnes.

Palladium (Pd) (46) is the least dense of the PGE metals and has the lowest melting point. It is so malleable that it can be beaten into extremely thin sheets. Its most important applications are in automobile catalytic converters, fuel cells, and in the tiny ceramic capacitors used in computers and cell phones. It was discovered in platinum ore in 1802 by William Wollaston.

World production in 2012 was 201 tonnes, mainly from Russia and South Africa (table 6.1D). Because of its industrial importance, it is considered to be a strategic mineral and its supply is tied to that of platinum; however, much of it can be recycled.

Osmium (Os) (76) is the densest element and the hardest metal. It usually occurs as an alloy with iridium

country	amount in tonnes
China	403
Australia	250
USA	235
Russia	218
Peru	161
South Africa	160
Canada	104
Mexico	97

A. Gold

country	amount in kilo-tonnes
Mexico	5.36
China	3.9
Peru	3.48
Australia	1.73
Russia	1.5
Bolivia	1.21
Chile	1.15
Poland	1.15

B. Silver

country	amount in kilo-tonnes
South Africa	1.33
Russia	24
Zimbabwe	11
Canada	7

C. Platinum

country	amount in tonnes
Russia	82
South Africa	74
USA	12.3
Canada	12.2
Zimbabwe	9

D. Palladium

country	amount in kilotonnes
Chile	13.2
Australia	12.8
China	4.5
Argentina	2.7

E. Lithium

country	amount in kilotonnes
USA	11,100
Turkey	1,800
Kenya	500
Mexico	290

E. soda ash*

country	amount in kilotonnes
Canada	8,980
Russia	5,470
Belarus	4,760
China	4,100
Germany	3,120

G. Potassium (oxide)

country	amount in kilotonnes
China	698
Russia	29
Israel	27
Kazakhstan	21
Brazil	14

H. Magnesium

country	amount in kilotonnes
China	30-35
Russia	4
USA	1-2

I. Calcium

Table 6.1 Metal production by main producing countries – IV
Data sources: A–H, US Geological Survey (USGS) 2014; I, Minor Metals Trade Association, 2012.

and, like it, is produced as a by-product of nickel refining. Although osmium metal is not very reactive, when in powder form it readily oxidizes and is toxic when inhaled or ingested. Its main application is in the form of a platinum-osmium alloy in hard needles, instrument pivots and electrical contacts.

Osmium is present in minute quantities in the crust, estimated at around 0.0015ppm, and occurs with the other PGE metals; like them, it is produced as a by-product of the electro-refining of copper and zinc. Only between one and two tonnes are produced annually, mostly from South Africa.

Iridium (Ir) (77) is also extremely rare in the crust (*c.* 0.001ppm) but is abundant in meteorites, which explains the famous '**iridium anomaly**' – a thin layer of sedimentary rock believed to represent the fallout from a giant meteor impact 65 million years ago. This is thought to have caused the end-Cretaceous mass extinction that contributed to the demise of the Dinosaurs (and much else besides). The element was discovered in 1803 by Smithson Tennant along with osmium, by dissolving platinum in very strong acid.

Pure iridium is a brittle, lustrous and silvery metal, found alloyed with osmium and produced as a by-product of platinum refining. It is used in a variety of specialized products including the tips of spark plugs and high-temperature crucibles, where its high melting point (2466°C) is critical.

World production of iridium is currently only about 3 tonnes, from the same sources as platinum. However, it is considered to be an important strategic metal because of a projected expansion of several niche uses, although the high cost of production is likely to limit any pressure on future resources.

Rhodium and Ruthenium
The metals rhodium and ruthenium are both extremely rare and share many properties with the platinum group metals with which they are sometimes grouped.

Rhodium (Rh) (45) is considered to be one of the most precious metals.

It is used as a coating in jewellery because of its lustre and reflectivity. In 1979, a special rhodium-plated disc was presented to Sir Paul McCartney to commemorate his achievements. An estimated 30 tonnes of the metal are produced annually, from the same sources as the PGE: 80% of it from South Africa.

Ruthenium (Ru) (44), being hard and unaffected by exposure to air, water or acids, is also used as a surface coating for jewellery, and alloyed with platinum and palladium in catalysts, electrical contacts, and as a coating for computer hard discs. Like these metals, it is produced as a by-product of nickel refining. World production is estimated at around 20 tonnes annually, from the same sources as rhodium. World reserves have been estimated at about 5 kilotonnes. Like the PGE, much can be recycled.

The alkali metals

The alkali metals are a group of six soft, highly reactive metals that occupy the extreme left side of the Periodic Table (*see* Fig. 2.2). They dissolve easily in water to form alkaline solutions.

Lithium (Li) (3) is the lightest of the metals and is so highly reactive that it has to be stored with a coating of petroleum jelly to prevent it from combining with air or water. Lithium first achieved widespread notice in 1949, when it was discovered that small doses of lithium carbonate can be effective in treating patients with mental health problems, replacing the previous practice of electro-convulsive therapy, and it is now widely used for this purpose. However, it is highly toxic if the recommended dosages are exceeded. The metal was first discovered by Johan Arfvedson in 1817.

Lithium is comparatively abundant in the crust – estimated at *c.*20ppm.The main ores of lithium are **spodumene**, **petalite** (both lithium-aluminium silicates) and **lepidolite** (a complex lithium mica) found in granitic pegmatites. Large deposits of spodumene occur in South Africa, Russia, China, Zimbabwe and Brazil; lithium can also be recovered from lake brines and sedimentary clays. It can be extracted electrolytically from the chloride.

Apart from its medical use, lithium oxide is employed in heat-resistant glass, and lithium-ion batteries are a familiar feature of wrist watches and cameras. Combined with aluminium, lithium also provides a more lightweight material for aircraft construction.

World production in 2012 was estimated at 35 kilotonnes, with the bulk provided by Chile, Australia, China and Argentina (table 6.1E). In the same year, reserves were estimated at 13 million tonnes. Increasing use of lithium batteries for electric and hybrid cars means that the consumption of lithium is likely to rise considerably in the future; however, there would seem to be adequate resources and there are a number of alternative sources, including sea water.

Sodium (Na) (11) is best known because of its chloride, **halite** (common salt); however, the symbol Na comes from the Latin *natrium*, for sodium hydroxide. Like lithium, sodium is highly reactive, burning with a bright yellow flame if dropped into water – a favourite school chemistry demonstration. The reaction with water produces sodium hydroxide and hydrogen, which burns in the presence of oxygen.

Salt is essential for human life, and has been valued highly, traded, and taxed since the earliest recorded times. Its importance is preserved in familiar phrases such as 'worth his salt' and 'salt of the Earth'. Our salty blood is the mammalian equivalent of the seas that our remote ancestors inhabited. It is essential to the human body because of its role as an electrolyte, employed in the transmission of nerve impulses and the regulation of bodily fluid levels. However, excessive salt intake is harmful, leading to high blood pressure, and much government publicity is aimed at reducing the overuse of salt in processed food. Sodium compounds are used in a number of manufacturing processes including the production of glass and paper, and the hydroxide is an important alkali in the chemical industry. Sodium metal for industrial purposes is extracted electrolytically from the molten chloride.

Sodium is comparatively abundant, making up about 2.3% of the crust and 0.04% of sea water. The main source of halite (sodium chloride) is from rock-salt deposits, which are relatively common in the geological record, and from the evaporation of sea or salt-lake water, and these are exploited in large quantities (see chapter 8). Sodium for industrial purposes is mostly produced in the form of **soda ash** (sodium carbonate) or sodium sulphate, both from evaporite deposits. 52,900 kilotonnes of soda ash and *c.*6 million tonnes of sodium sulphate were produced in 2012. The USA was the major producer of soda ash (table 6.1F); Mexico, Spain, the USA and Canada were all major suppliers of sodium sulphate.

However, only around 100 kilotonnes of sodium metal are produced annually. Because of the quantities of sodium theoretically available in sea water, resources are essentially inexhaustible.

Potassium (K) (19) is even more reactive than sodium, exploding in contact with water and burning with a lilac-coloured flame. Potassium salts, termed **potash**, (e.g. chloride, sulphate and nitrate), have been known since ancient times and have been used to flavour food, preserve meat and improve soil. Potassium was originally obtained by dissolving wood ash and precipitating the potash as a white powder. The symbol K is derived from the Latin *kallium* for potash. However, potassium was not recognized as a separate metal until 1807 when it was isolated by Sir Humphry Davy, the famous British chemist who invented the 'Davy' miners' lamp.

The discovery in 1840 by Justus von Leibig that potassium is essential for plant growth led to the development of fertilizers containing potassium, along with nitrogen and phosphorus, which revolutionized agricultural production.

Potassium is comparatively abundant, making up about 2.1% of the crust. It occurs in several silicate minerals, especially **orthoclase feldspar**, which contains *c*.5% potassium and is an important constituent of granite. It is easily leached out by acid rainwater, finding its way into the sea, where it makes up *c*.400ppm. It occurs widely in evaporite deposits as sulphates and chlorides. Some of the most productive deposits are the **Zechstein** (mid-Permian) deposits that underlie the North European Plain, extending from Eastern Britain across the North Sea

and Germany to Poland. The potassium hydroxide is isolated from the accompanying sodium and magnesium salts by means of their differing solubilities, and the metal is separated from the hydroxide by electrolysis.

Potassium, like sodium, is vital for human life because of its role as an electrolyte in regulating fluid balance between the cells and their surrounding fluid. Nerve cells are particularly affected by potassium deficiency, which can cause various effects such as muscle cramp. Physical activity causing excessive perspiration results in the loss of potassium as well as sodium salts, and these should be replaced by drinking liquids containing both salts in solution. Also like sodium, excess potassium is harmful – potassium chloride has been employed medically in euthanasia.

Around 95% of the potassium produced is destined for fertilizers in the form of chlorides, sulphates and nitrates, but the remaining 5% has a wide variety of functions. For example, potassium nitrate is the main ingredient of gunpowder; potash is used in the manufacture of glass, soap, fluorescent lamps and dyes; potassium bisulphite is a food preservative; potassium chlorate is required for explosives, and potassium cyanide to dissolve precious metals (see the section on gold); potassium chromate gives the orange sparkle in fireworks.

Total world production of potash was 32.7 million tonnes in 2012, the top five producers being Canada, Russia, Belarus, China and Germany (table 6.1G). Reserves were estimated at about 6000 million tonnes, but in view of its abundance, theoretical

resources are effectively unlimited.

Rubidium (Rb) (37) has very similar properties to the other alkali metals and is so reactive in air or water that it has to be stored in a glass container filled with inert gas. It was discovered in 1861 by Robert Bunsen and Gustav Kirchhoff by spectroscopy. It is comparatively abundant in the crust (*c*.90ppm) and occurs in several potassium-bearing silicate minerals such as orthoclase feldspar and the mica lepidolite, and also as a hydroxide in evaporite deposits. Lepidolite, which occurs in granitic pegmatites, contains up to 3.5% of rubidium and is an important source.

Rubidium is well known to geologists in two ways: firstly, the rubidium/strontium ratio is a good indicator of the degree of fractionation in igneous rocks; and secondly, a widely used method of dating old rocks involves the radioactive isotope rubidium-87, which has a half-life of 50 billion years and decays to strontium-87, so that the (87Rb/87Sr) ratio reflects the period of time since the rock first crystallized, provided it has not suffered any disturbance in subsequent events.

There are few industrial uses for rubidium and only about 2–4 tonnes are produced annually, mainly from Canada and Italy. The metal is sourced from sulphate deposits in evaporites and separated by fractional crystallization. Niche uses include highly accurate atomic clocks and, in the form of its radioactive isotope Rb-82, in emission tomography.

Caesium (Cs) (55) is even more reactive than rubidium and has to be stored in the same way. It melts at 28°C, just above room temperature. It was also discovered by Robert Bunsen and Gustav Kirchhoff by spectroscopy in

1860. The radioactive isotope caesium-137 is notorious because of its release as a result of the Chernobyl nuclear explosion; the caesium dust in the atmosphere was washed into the soil and absorbed by plants. Large numbers of livestock throughout Western Europe were affected: in some areas, such as Wales, livestock could not be sold for food for up to 25 years afterwards.

Like rubidium, caesium is used in atomic clocks, which have an accuracy of 1 second in 300,000 years! The international standard second is defined on the basis of a specific number of cycles of radiation emitted, or absorbed, by the caesium atom. However, the main industrial application is in oil drilling fluid, in the form of caesium formate, because of its high density and low reactivity. Other applications include photo-electric cells, optical components of infra-red spectrometers, industrial gauges and, as the radioactive isotope Cs-137, for medical purposes.

The crustal abundance of caesium is about 3ppm. As one of the incompatible elements (see chapter 3), it is concentrated in the late-stage phases of acid magmas in pegmatites, and also in evaporites in the form of chlorides and hydroxides. However, the main source is **pollucite** (caesium aluminium silicate). Annual world production is only 5–10 tonnes, principally from Canada, which contributes two-thirds of the world's supply, and has estimated reserves of *c.*350 kilotonnes – more than enough to supply any likely future demand.

Francium (Fr) (87) is the heaviest of the alkali metals and is one of the least stable elements in the Periodic Table. It is intensely radioactive and occurs only in tiny quantities in uranium ores as a product of the radioactive decay of **actinium** (see chapter 7). Its longest-living isotope, francium-223, has a half-life of 22 minutes; it has been estimated that at any one time, only about 30 grams of the element are in existence in the Earth.

The element was discovered in 1939 by Marguerite Perey at the Curie Institute in Paris. It is now created only for research purposes and has no commercial applications.

The alkaline earth metals

The six alkaline earth metals occupy the column next to the alkali metals at the left-hand side of the Periodic Table. They are all solid and metallic at room temperature. The term 'alkaline earth' refers to the fact that their oxides are strongly alkaline in water.

Beryllium (Be) (4) is the fourth lightest metal; it is also strong, resistant to corrosion, and has a very high melting point. It was first isolated in 1828 and named after the mineral **beryl** (beryllium-aluminium silicate), a gemstone whose crystals can reach up to six metres in length! The green variety of beryl is known as **emerald** (see section on chromium in chapter 5).

Beryllium has few industrial applications; a few percent added to copper produces a non-sparking, high-strength metal ideal for cutting tools in oil wells, where there is a danger of combustion. It also has the unusual property of being able to reflect neutrons, but is transparent to electrons, which led to its use in nuclear warheads. It has no known biological role but is toxic to humans.

Crustal abundance of beryllium is around 2ppm and it makes up about 6ppm in soils. It occurs in the form of both oxides and silicates; however, the main industrial sources are beryl, which is found in a number of countries, and the hydrated beryllium silicate, **bertrandite**, which is mined in the USA. Total world production in 2012 was 240 tonnes, and only the USA and China produced significant quantities. Total world reserves have been estimated at *c.*400 kilotonnes, and much of the metal is currently recycled.

Magnesium (Mg) (12) is the fourth most abundant element in the Earth, making up 13% of its mass, and magnesium silicates such as **olivine** are one of the main constituents of the mantle. Magnesium metal is highly reactive and burns with an intense white flame, hence its use in marine flares. **Epsom salts** (magnesium sulphate) have been used since the seventeenth century as a laxative and '**milk of magnesia**' (magnesium hydroxide) for indigestion. However, the metal was not isolated until 1808, when Sir Humphry Davy obtained it by the electrolysis of sea water; because it is so easily soluble, sea water contains about 1290ppm.

Magnesium is a vital ingredient for almost all living organisms; it is an essential component of **chlorophyll**, which is the organic substance used by plants for photosynthesis. Plants consequently need to absorb magnesium from the soil, and magnesium deficiency causes leaves to turn yellow. Magnesium is also an essential trace element for the human body, especially for bone structure and for regulating blood sugar levels. The most important industrial application is in the production of magnesium-aluminium alloys, which are lighter and stronger than pure aluminium, for the air-craft and automobile industries.

Crustal abundance of magnesium is about 2.33%, and it is present in a wide variety of minerals, such as **dolomite** (calcium-magnesium carbonate), **magnesite** (magnesium carbonate), **brucite** (magnesium hydroxide), and **carnallite** (hydrated potassium-magnesium chloride) as well as many of the common silicates such as olivine and pyroxene. The main sources of magnesium for industrial production are dolomite and magnesite, which are both widely available as sedimentary deposits.

World production of magnesium in 2012 was 802 kilotonnes, the top five producers being China, Russia, Turkey, Austria and Slovakia (table 6.1H). There are very large reserves both of carbonates and brine deposits, and the theoretical supply of magnesium from all the possible sources is effectively unlimited.

Calcium (Ca) (20) is the fifth most abundant element in the Earth's crust – even more abundant than magnesium. It occurs in a wide variety of igneous and sedimentary rocks and performs a number of functions essential to most forms of life. The metal is highly unstable, though not as reactive as magnesium, combining with air to form both hydroxide and carbonate. In water, it gives off hydrogen gas.

Calcium is an essential component of the vertebrate skeleton in the form of calcium phosphate, and needs to be constantly renewed to keep bone in good condition. Healthy adults require about 1500 milligrams per day. It is present in dairy products and vegetables. Calcium in the form of calcium carbonate is also vital to make the shells of marine animals such as molluscs, and is present in sea water in appreciable amounts.

Calcium oxide (**lime**, or **quicklime**) has been exploited since prehistoric times as mortar for building purposes. When water is added, lime becomes calcium hydroxide ('**slaked lime**', or **cement**) which gradually hardens in air to become calcium carbonate (i.e. limestone). **Plaster of Paris** (hydrated calcium sulphate) has been used to set broken bones since the tenth century. Lime is now employed industrially to make concrete for construction, for fertilizers to correct acid soils, in water treatment, and as a flux in steel manufacture. Calcium metal was first isolated by Sir Humphry Davy by electrolysis in 1808.

Calcium is present in abundance in the crust (*c.*4.15%) in the form of calcium carbonate (limestone and dolomite), together with other compounds such as **anhydrite** (calcium sulphate) and **gypsum** (hydrated calcium sulphate). Limestone and dolomite deposits are very common in the sedimentary record, and both gypsum and anhydrite occur widely in evaporite deposits. Most calcium is produced in the form of lime, whose resources are discussed in chapter 8. The bulk of calcium metal production comes from China, Russia and the USA (table 6.1I). The world supply of accessible calcium compounds (especially limestone) is effectively unlimited.

Strontium (Sr) (38) is another highly reactive metal, igniting spontaneously in air. It is named after the village of Strontian on the shores of Loch Sunart in western Scotland, where it was discovered in 1790 in carbonate ores from the local lead mine. Sir Humphry Davy isolated the metal from **strontianite** (strontium carbonate) in 1808.

Strontium was employed industrially in significant amounts from the late nineteenth to mid-twentieth century for producing sugar from sugar beet, and more recently for coating cathode-ray tubes. However, there are few applications now; strontium carbonate gives a brilliant red colour in fireworks, and the chloride has a niche use in toothpaste for sensitive teeth. The human body absorbs strontium in place of calcium with very similar effects, and small amounts can improve bone growth.

Strontium-90 is a highly radioactive isotope with a half-life of only 29 years, and was produced as a result of nuclear test explosions in the 1950s. Alarm at the high levels of Sr-90 in children's teeth during this period led eventually to the nuclear test ban treaty between the USA, Russia and the UK. 87Sr/86Sr ratios in teeth and bone help archaeologists to determine the areas of origin of skeletal material, since soils in different areas can be distinguished by their strontium isotope ratios and have a direct influence on human diets. The geological use of the 87Rb/87Sr ratio in rock dating has already been referred to.

Crustal abundance of strontium is about 370ppm, the main sources being the carbonate strontianite, from lead deposits, and the sulphate **celestine**, from evaporite deposits. About 228 kilotonnes were produced in 2012 from celestine, the main sources being China, Spain and Mexico (table 6.2A). Reserves were estimated at 6.8 billion tonnes.

Barium (Ba) (56) is chemically similar to the preceding alkali metals

country	amount in kilotonnes
China	100
Spain	80
Mexico	40.9

A. Strontium

country	amount in kilotonnes
China	4,200
India	1,700
Morocco	1,000
USA	666

B. Barites

country	amount in tonnes
China	100,000
Australia	3,200
India	2,900
Russia	2,400
USA	800

C. Lanthanides

Table 6.2 Production of strontium, barites and lanthanides by main producing countries
Data sources: US Geological Survey (USGS), 2014.

but is even more reactive in air, and does not occur naturally as a metal. Many of its compounds are abundant, especially barite (barium sulphate) which has been known since the 1600s, when it was discovered by an Italian alchemist that pebbles of barite, known as 'Bologna stones', when exposed to light during the day, would glow at night. The metal was isolated, along with the other alkali metals, by Sir Humphry Davy in 1808 – clearly a busy year for Sir Humphry!

Barite is widely employed in the oil industry in drilling fluid because of its high density, but perhaps its best known function is as a 'barium meal' for medical purposes, to diagnose gastric and intestinal disorders; it absorbs X-rays, thus enabling it to show up in scans. Barium carbonate is used in glass-making and as a rat poison; it also produces the green colour in firework displays. Barium metal is used to remove gases from vacuum tubes and as an alloy with various metals including steel.

Crustal abundance of barium is estimated at *c.*425ppm. It occurs mainly in the form of barite and **witherite**

(barium carbonate) in SEDEX lead-zinc ore deposits (see Figure 4.2A). Total world production of barite in 2012 was 9,200 kilotonnes, mainly from China, India, the USA and Morocco (table 6.2B). Reserves of barite were estimated at 360 million tonnes.

Radium (Ra) (88) is the heaviest of all the alkaline earth metals and is highly radioactive. It was made famous by the discovery by Marie and Pierre Curie in 1898 that the compound radium chloride, which they had extracted from uraninite (see below), emitted energy (α- and β-particles and γ-rays) causing their samples to glow with a bluish luminescence. The metal was isolated several years later. After its discovery, radium was exploited widely as a 'wonder cure' until its dangerous side effects were realized.

In more recent times, the main industrial application has been in luminescent paint for watches, aircraft instrument panels, etc. However, because of the problems associated with exposure to radioactivity, radium has been replaced by safer alternatives for most purposes, and its applications are now restricted to limited medical use in

cancer treatment. The extreme toxicity of radium is due to its ability to substitute for calcium in the human body, which means that it can be absorbed into bone cells, causing cancer.

Radium is present in minute quantities (*c.*0.14ppm) in all uranium ores as a breakdown product of uranium, of which the main ore is **uraninite** (uranium oxide), usually known as **pitchblende** (see chapter 7). The supply of radium therefore comes from uranium production, together with waste products from nuclear reactors. Because of the short half-lives of its radioactive isotopes, all the radium now present in the Earth is, in geological terms, of recent origin.

The lanthanides (rare-earth elements)

There are 15 lanthanide elements, with atomic numbers from 57 to 71, placed between barium and hafnium in the Periodic Table. However, they are usually displayed in a separate row beneath the transition metals (see Figure 2.2). They are usually known by geologists as the '**rare-earth elements**', but some of them are not particularly rare (lanthanum itself is three times more common than lead); nor are they 'earths' but normal metals; however, they are found in less common minerals, and are very difficult to separate from each other because they all have similar properties. The elements with odd atomic numbers are much rarer than those with even numbers because they have a less stable atomic structure. They are generally soft, but increase in hardness along the series, and are good electrical conductors. Small differences in solubility are used to separate them by fractional crystallization. Yttrium and scandium (see chapter 5) are sometimes also regarded as rare earths although they belong to the transition metals rather than the lanthanides.

The lanthanide series may be divided into lighter and heavier: the lighter (lanthanum to europium) are concentrated in the crust and the heavier (gadolinium to lutetium) in the mantle; the relative proportions of light to heavy are used by geologists to distinguish magmas from different sources. Several have radioactive isotopes, and the samarium-neodymium ratio is used to date rocks.

Lanthanides are frequently employed as alloys in the same ratios as they occur in nature: typically this is 50% cerium, 25% lanthanum, 15% neodymium and the rest, 10%. Alloys with iron, cobalt or magnesium are used as catalysts in various specialized industrial processes. They have the capacity of storing large quantities of hydrogen, which makes them ideal for storage batteries: for example, the nickel-hydride batteries in hybrid automobiles. The typical hybrid car battery contains 10–15 kilograms of lanthanide. About 18 kilotonnes are used annually, mainly in catalytic converters, petroleum refining, and as permanent magnets and glass polishing.

The principal ores containing the lanthanide elements are **monazite** (cerium-lanthanum-yttrium phosphate) and **bastnäsite** (cerium-lanthanum-yttrium fluorocarbonate). Monazite occurs in monazite-rich heavy-mineral sands. Monazite is an accessory mineral in granites and pegmatites, and is concentrated in heavy-mineral sands – a type of placer deposit (see chapter 4). Bastnäsite occurs originally in alkali granites and pegmatites,

and in carbonatites, but the mineable resources are in sediments derived from these. Monazite sands usually contain all the lanthanides, whereas bastnäsite lacks the heavier elements. Most mined lanthanide ores are currently produced in China from bastnäsite, but considerable lanthanide reserves are situated in Australia, South Africa, Brazil, Malaysia, India, and the USA, in both monazite and bastnäsite. World production in 2012 was 110 kilotonnes, leading producers being China, Australia, India and Russia (table 6.2C). World reserves were estimated at about 110 million tonnes.

Lanthanum (La) (57) was discovered in 1839 by Carl Mosander in cerium nitrate but the element itself was not isolated until 1923. Its crustal abundance is about 32ppm, the main source being monazite sands; however, because of the difficulty of separating it from the other lanthanides, it is rarely used alone. World production is about 12.5 kilotonnes annually.

Cerium (Ce) (58) was isolated much earlier than lanthanum, in 1875, by Swedish scientists Jöns Berzelius and Wilhelm Hisinger from the cerium salt, **cerite** (cerium silicate). It is the most abundant of the lanthanides, making up about 67ppm of the crust. It is an environmentally 'friendly' element: adding it to diesel fuel improves burning efficiency, causing less harmful emissions, and cerium sulphide gives a red pigment that has replaced more toxic alternatives. It can also be used in firelighters – igniting spontaneously in air. World production is about 24 kilotonnes annually.

Praseodymium (Pr) (59) was isolated in 1885 by Carl von Welsbach. It is relatively abundant, making up *c.*9ppm of

the crust, and amounts to around 5% of the lanthanide content of the lanthanide ores. It has a number of industrial applications: when alloyed with magnesium, it forms a super-strong alloy for aircraft engine components; added to glass, it filters out infra-red radiation and is therefore used in welders' masks. When a magnetic field is applied to a sample of praseodymium, its temperature is lowered, and scientists have been able to cool it to within one-thousandth of a degree of absolute zero! About 2.5 kilotonnes are produced annually.

Neodymium (Nd) (60) is the third most abundant of the lanthanides, constituting between 10% and 18% of the lanthanides extracted from the ores. Its crustal abundance is about 42ppm. It was isolated in 1885 by Carl von Welsbach along with praseodymium. It also has many important industrial applications. Its magnetic properties, when alloyed with iron and boron, are better than any other known permanent magnet, prompting its use in small powerful magnets for various purposes including earpieces, computer hard-drives and hybrid cars. It is also used to make artificial neodymium-aluminium-garnet crystals for lasers. About 7 kilotonnes of neodymium oxide are produced annually.

Promethium (Pm) (61) is so highly radioactive that the naturally occurring element is almost non-existent. The most stable isotope has a half-life of only 17 years, so that none of the original element still exists. It was only discovered in 1945 in the debris from a nuclear reactor as one of the decay products of uranium.

Samarium (Sm) (62) was discovered in 1879 by Paul Emile Lecoq de Boisbaudran. It makes up 7ppm of the crust and about 3% of monazite ore. One of its main industrial applications is in the manufacture of samarium-cobalt magnets, which retain their magnetism at very high temperatures (more than 700°C). The isotope samarium-149 absorbs neutrons and is used as a moderator in the control rods of nuclear reactors. About 700 tonnes are produced annually.

Europium (Eu) (63) is one of the rarer lanthanides, making up about 2ppm of the crust. It was isolated in 1901 by Eugène-Anatole Demarcay. Europium is well known to geologists because of the so-called 'europium anomaly'; this relates to the fact that the europium concentration in most minerals is either anomalously high or anomalously low, compared to the other lanthanides; this feature enables igneous rocks from different sources to be distinguished. Europium was formerly added to the red phosphors as a coating for the inside of cathode-ray screens because of its luminescent property; however, its main functions now are in superconductors and low-energy light bulbs. Only about 100 tonnes are produced annually.

Gadolinium (Gd) (64) is one of the commoner lanthanides, present in amounts of about 6.2ppm in the crust. It was first isolated in 1886 by Paul de Boisbaudran. Gadolinium is one of the best neutron absorbers, and for this reason is employed in nuclear reactor control rods. This property also accounts for its use in magnetic resonance imaging (MRI) scanning: gadolinium injected into a particular part of a patient shows up on the scanner screen. Gadolinium

is also strongly magnetic up to 19°C, when it suddenly loses its magnetism, enabling it to be used in magnetically controlled refrigeration systems. About 400 tonnes are produced annually.

Terbium (Tb) (65) was discovered and isolated in 1843 by Carl Mosander, along with several other lanthanides (yttrium, ytterbium and erbium) in the mineral yttrium oxide, from the village of Ytterby in Sweden. It is one of the rarer lanthanides, making up c.1.2ppm of the crust: monazite contains only 0.03% of it and bastnäsite 0.02%, although certain clays in China contain as much as 1%. The element has a number of important niche industrial applications, however. The terbium alloy Terfenol-D™ has the unusual property of expanding and contracting in response to a magnetic field. This has enabled an electric current to transmit variations in frequency to a solid object capable of resonating, thus enabling sound to be amplified. It also features in naval sonar equipment, and in green phosphors for various electronic devices. About 10 tonnes are produced annually.

Dysprosium (Dy) (66) is the most abundant of the heavy lanthanides, forming 5.2ppm of the crust. It was discovered in 1886 by Paul de Boisbaudran, but not finally isolated until the 1950s by Frank Spedding, using ion-exchange chromatography. In view of the difficulty in extracting this element, the name 'dysprosium', given to it by De Boisbaudran (after the Greek word *dysprositos* meaning 'hard-to-get'), was indeed appropriate. Dysprosium, as a component of a niobium-iron-boron alloy, plays a very important role in the powerful

magnets in high-efficiency electric motors such as those used in hybrid electric cars. Every hybrid car needs about 100 grams of dysprosium, so that global requirements for the metal are expected to rise steeply in the future, and the US Department of Energy anticipates a global shortage by 2015! About 100 tonnes of dysprosium are produced annually at present, almost all from the yttrium-bearing clays of southern China. Dysprosium is also present in the principal lanthanide ores in proportions of 7–8%; however, the difficulty of extraction would increase the cost of production.

Holmium (Ho) (67) was discovered in 1878, and subsequently isolated and named by Swedish chemist Per Cleve after the Swedish capital, Stockholm. Holmium is one of the rarer lanthanides, with a crustal abundance of about 1.3ppm. It has the highest magnetic strength of all the elements, and this property is exploited in a number of specialist roles, including MRI scanning. It is also added to yttrium-iron garnets for the lasers used in surgical and dental procedures. About 10 tonnes are produced annually, mainly from the Chinese yttrium-bearing clay deposits. It is estimated that about 400 kilotonnes of reserves are available from the known lanthanide ore deposits.

Erbium (Er) (68) is another of the lanthanide elements discovered and named after the Swedish village of Ytterby by Carl Mosander in 1843. However, it was not isolated until 1934. It is one of the more common lanthanides, present in the crust in proportions of about 3.5ppm. An important industrial application is in fibre-optic communications: when erbium is added to the glass fibres, they assume laser-like properties, resulting in the amplification of the original pulse of light. Erbium is also used to make pink glass – and hence rose-tinted spectacles! About 1181 tonnes were produced in 2014.

Thulium (Tm) (69) is one of the rarer lanthanides, with a crustal abundance of c.0.5ppm and comprises only 0.07% of monazite. It was discovered in 1879 by Per Cleve and named from the ancient word *Thule* meaning 'land at the end of the known world', referring to the difficulty of extracting it. It has no particularly unique properties and is so difficult and expensive to extract that it has found few applications. Lighting engineers use it to produce the green colour in arc lamps and it has also found a role in yttrium-aluminium garnet lasers. About 50 tonnes of thulium oxide are produced annually.

Ytterbium (Yb) (70) was discovered in 1878 by Jean Charles Galissard de Marignac from the yttrium ore in Ytterby, but the pure metal was not isolated until the 1950s by the ion-exchange technique. The synthetic isotope ytterbium-169 is a strong emitter of X-rays and is exploited in portable X-ray devices for emergency medical applications and also for diagnosing faults in machinery. It is added to stainless steel to improve strength, and can also improve the quality of laser light. It also has the unusual property of behaving as a conductor up to 16,000 atmospheres pressure, when it becomes a semiconductor, but by 39,000 atmospheres becomes a normal conductor again. This property has enabled it to be used in pressure gauges in extreme conditions such as those in nuclear explosions.

Crustal abundance is about 3ppm. The Chinese yttrium clays contain 3–4% of ytterbium oxide. About 50 tonnes are produced annually.

Lutetium (Lu) (71) is the last element in the lanthanide series, and is the hardest and most dense. It is regarded by some as a transition metal rather than a lanthanide. It was discovered in 1907 by Georges Urbain – the last in the series to be found. It has few applications and is very expensive to produce. It is used in positron emission tomography for medical imaging. At one time the price reached $75,000 per kilogram but has subsequently decreased. The crustal abundance of lutetium is about 0.5ppm, and it is sourced from the principal lanthanide ores. Annual production is only about 10 tonnes.

7 Resources of non-metallic elements

The non-metallic elements consist of hydrogen, the so-called 'noble gases', the halogens and a group called simply 'other non-metals' in the Periodic Table, which include important elements such as carbon, nitrogen and oxygen (*see* Fig. 2.2).

Hydrogen

Hydrogen (H) (1) is the first element in the Periodic Table and has the simplest atomic structure, with only a single proton and one electron. It was one of the first three elements to be formed in the 'Big Bang', along with helium and lithium. Hydrogen is the 'fuel' that keeps the Sun and other stars burning, due to the nuclear reaction that converts it to helium with the release of enormous amounts of energy. The hydrogen bomb employs the same principle: an outer nuclear fission bomb generates sufficiently high temperatures to detonate the inner nuclear fusion bomb by forcing hydrogen atoms to combine under very high temperatures and pressures.

Hydrogen is a gas at standard temperatures and pressures; it is highly combustible, combining with oxygen to produce water. This was first recognized in 1766 by Henry Cavendish, who proved that water was not, as previously thought, one of the 'basic elements'. Hydrogen is the basis of the complex hydrocarbons that are an essential component of all living things.

The main industrial uses for hydrogen currently are in fertilizer manufacture, through combining it with nitrogen to make ammonia, and in petroleum refining. It is produced either from natural gas or by electrolysis of water. Over 50 million tonnes are produced annually, much of it on the same site as it is used; the USA presently accounts for a large proportion of the production. Other important sources are Brazil, Germany, India and China.

However, the future role of hydrogen undoubtedly lies in the hydrogen fuel cell, which is a much more environmentally acceptable source of energy than the hydrocarbons, since the main waste product is water. Supplies of hydrogen from water are effectively unlimited (crustal abundance is about 1.4%) and the main factor affecting its industrial applications is the cost of producing, then storing, it (*see* the section on the lanthanides in chapter 6).

The metalloids

The metalloids are a group of elements, the best known being silicon, that occupy a diagonal strip in the Periodic Table to the right of the transition metals. The metalloids have some metal-like characteristics such as good electrical conductivity.

Boron (Bo) (5) is chiefly known because of its compound **borax** (hydrated sodium borate) which has been used for millennia as a detergent, a fungicide and a pesticide. It was discovered in Tibet and traded on the 'Silk Road' to Europe in the Middle Ages. In later centuries it became a cosmetic. The metal itself was not isolated until 1892. In more recent times it has been widely used as an abrasive. Boron, albeit in only trace amounts, is essential to both crops and animals, and boron deficiency is a recognized agricultural problem.

The crustal abundance of boron is about 10ppm; it is only found in the form of boro-silicates, such as **tourmaline** and **axinite**, but the main source of borate minerals is evaporite deposits of **colemanite** (hydrated calcium borate) from saline lakes and volcanic hot springs. World production of boron in 2012 was 4420 kilotonnes, the main producers being Turkey, the USA, Argentina, Chile and Russia (table 7.1A).

Silicon (Si) (14) is the second most abundant element in the crust, after oxygen. The name is derived from the Latin *silicis* for **flint**, one of the first minerals to be used by early humans. Silicon is best known, however, in the form of **silica** (silicon dioxide), which forms the mineral quartz – the pure form of ordinary beach sand, and the dominant component of most sandstones. Quartz crystals (Fig. 7.1) are commonly found in nature, and many varieties, such as **amethyst**, are considered to be gemstones. Silicon was discovered by Jöns Berzelius in 1824 by heating potassium fluorosilicate with potassium metal.

One of the earliest uses for silica was

country	amount in kilotonnes
Turkey	2,500
Argentina	650
Chile	440
Russia	400
USA	W

A. Boron

country	amount in kilotonnes
China	5,050
Russia	733
USA	383
Norway	339
Brazil	225

B. Silicon

country	amount in kilograms
China	105,000
Russia	5,000
USA	W

C. Germanium

country	amount in kilotonnes
China	26
Chile	10
Morocco	8
Russia	1.5
Belgium	1.0

D. Arsenic

country	amount in kilotonnes
China	145
Russia	6.5
Bolivia	4.0
South Africa	3.8
Tajikistan	2.0

E. Antimony (production)

country	amount in kilotonnes
China	950
Russia	350
Bolivia	310
Tajikistan	50

F. Antimony (reserves)

country	amount in tonnes
Japan	45
Russia	35
Canada	11
USA	W

G. Tellurium

country	amount in kilotonnes
China	800
India	160
Brazil	110
North Korea	30
Canada	25

H. Carbon (graphite)

country	million carats
Botswana	20
D.R Congo	17
Russia	15
Australia	8
South Africa	4

I. Diamond

Table 7.1 Production of non-metals by main producing countries – I
Data sources: US Geological Survey (USGS): A, B, D–H, 2014; C, 2015; I, Diamonds.net, 2012.
Note: 1 carat = 200mg. W, data withheld.

in the production of glass. However, its most important application now, when alloyed with small amounts of other elements, is in the manufacture of microchips. These act as tiny transistors (switches) to control electric currents, and are vital to the functioning of all electronic devices, especially computers. Other modern uses include the manufacture of alloys: with aluminium to improve casting quality, and with carbon as an abrasive. Pure quartz crystals resonate with a specific frequency, and are employed in that capacity in clocks and watches. Silicone is a polymer based on silicon and oxygen, which has many commercial applications including,

notoriously, breast implants; these, in many cases, broke down within the body and were feared to be toxic. However, silicon in small quantities is now believed to be relatively safe, and silicon gels are used directly on the skin to help to heal scar tissue.

World production of silicon in 2012 totalled 7700 kilotonnes, the five top producers being China, Russia, the USA, Norway and Brazil (table 7.1B). Because of the widespread abundance of sand and sandstone (see chapter 8), reserves of silicon are virtually unlimited.

Germanium (Ge) (32) was discovered in 1886 by German chemist Clemens Winkler, who named it after his country. The first **semi-conductors** were made from germanium oxide, although this has now been largely superseded by the much cheaper silicon. The main destinations now for germanium oxide include solar panels and fibre-optic cables, where its high refractive index and low optical dispersion make it particularly suitable. It is also completely transparent to infra-red light, so is ideal for night-vision goggles.

Crustal abundance of germanium is 1.5ppm and there are few minerals that contain appreciable quantities of it. The zinc ore sphalerite contains up to 0.3% but the main sources are silver-lead-copper ores; it is also recovered from the ash residue of certain coal-fired power plants. The total world production in 2012 was 150,000 kilograms, of which most came from China, Russia and the USA (table 7.1C).

Arsenic (As) (33) is perhaps best known as the poison of choice for the thriller writers, and is believed to

Figure 7.1 Quartz crystals
Image credit: © Shutterstock, Madlen.

have (accidentally) caused the death of Napoleon Bonaparte through the dye on his wallpaper! Once routinely used, and readily available, as a weed-killer, it has now fallen out of fashion because of its toxicity – just 100mg are enough to cause death, although small amounts are tolerated by the human body. In its more orthodox applications, it is alloyed with certain metals (e.g. gallium) in semi-conductors to improve their effectiveness.

Crustal abundance of arsenic is only about 1.8ppm, but it occurs widely as a sulphide (**arsenopyrite**) along with other sulphide ores, and most of the arsenic produced commercially is a by-product of copper, lead, cobalt or gold mining. World production of arsenic in 2012 totalled 46.7 kilotonnes, mostly from China, Chile and Morocco (table 7.1D). There are large reserves.

Antimony (Sb) (51) is rarely found in its pure state, but the compound **stibnite** (antimony sulphide) occurs widely, and has been known for many centuries; the symbol Sb comes from the Latin *stibnium*. It is thought that the Byzantine navies used it in combination with crude oil and **saltpetre** to create 'Greek fire'; this was deployed against enemy ships because it gives off copious amounts of heat when ignited, and keeps burning even when floating on water. Kohl, used by the ancient Egyptians as an eye cosmetic, is black antimony sulphide.

Antimony is toxic, although not to the same degree as arsenic; nevertheless, antimony compounds have been extensively used for medical purposes, including burns treatment, and as an emetic. Mozart is said to have been poisoned by an accidental dose of antimony tartrate. It was first isolated and named by Vannoccio Biringuccio in 1540.

In modern applications, antimony is alloyed with tin and lead in the manufacture of solders, bullets and bearings. Antimony chlorides and bromides are used in fire retardant materials and, because it reflects infra-red radiation, in military camouflage paint.

The crustal abundance of antimony is only 0.2ppm but it occurs widely in many different minerals, although stibnite is the commonest. It is found associated with both volcanic and metamorphic rocks, often in quartz veins. World production of antimony in 2012 was about 174 kilotonnes, most of it from China (table 7.1E). There is concern over the global resources of this mineral, since it is feared that the known reserves (table 7.1F) will become exhausted in a few years at the present rate of production. Antimony is listed first in the Risk List published by the British Geological Society in 2011 (see table 12.1) due to the lack of significant supplies outside China, whose reserves are being rapidly depleted.

Tellurium (Te) (52) was discovered by Franz-Joseph Müller von Reichenstein in 1783 from a sample of gold telluride, but not isolated until 1798. The crustal abundance of tellurium is only 0.001ppm. It occurs rarely in the native state but mainly as the oxide (**tellurite**), and various other compounds along with gold and silver. However, most of the tellurium produced is obtained as a by-product of copper processing or from the electrolytic refining of sodium telluride solution.

Tellurium is employed industrially as an alloy with lead, copper and stainless steel to improve their quality, and to vulcanize rubber. However, its most important application at present is in the manufacture of memory chips and solar cells. The expansion of the solar power industry initially caused a steep rise in the price of tellurium, and concern about its sustainability, especially as copper production is declining; however, consumption and price have subsequently declined. Production of tellurium in 2012 from Japan, Russia, and Canada are listed in table 7.1G; US production figures are not released. Reserves were estimated by the US Geological Survey at about 24 kilotonnes.

Polonium (Po) (84) is highly radioactive; it gained notoriety in 2006 when Russian dissident and ex-KGB officer Alexander Litvinenko was poisoned by it and died a painful death shortly afterwards in London. Polonium is one of the deadliest poisons known –less than 0.1 microgram can be fatal. It was discovered in 1898 by Marie and Pierre Curie when studying the uranium ore, pitchblende, which contains a minute amount of the element (one part in 10^{10}) derived from the radioactive breakdown of the uranium. Polonium-210 is the most common isotope and is so radioactive that a lump of it will glow in air. It is rarely found outside the laboratory and is produced in a nuclear reactor by bombarding bismuth with neutrons.

Only around 100 grams are produced annually, almost all in Russia.

Polonium was once used as an initiator in the early atomic bombs and was subsequently deployed by the former USSR as an atomic heat source, for example in their Moon rover vehicles. More recently, it has been employed in various anti-static devices in which the air is ionized by the radiation, thus eliminating the static charge. Because of its short half-life, it has to be continually renewed. Its extreme toxicity makes it very dangerous to handle; since it occurs in tobacco, it is the most probable cause of cancers resulting from smoking.

The carbon group

The elements in this group are usually classified as 'other non-metals', which is not a very helpful description; they consist of the following: carbon, nitrogen, oxygen, phosphorus, sulphur and selenium. They display a wide variety of properties but, with the exception of carbon, are generally poor conductors of heat and electricity, have lower melting points and boiling points, and lower densities than most metals.

Carbon (C) (6) is one of the most important of the elements, since its organic chemical compounds are the basis of all forms of life. It helps to power the nuclear reactions in the Sun, and is an integral part of fossil fuels such as oil and coal. Carbon was known as far back as 3750BCE to the ancient Egyptians, who used it to process ores to make bronze. More recently, in the form of **coke**, it became an essential part of steel production.

The name 'carbon' comes from the Latin *carbo* for charcoal which, along

with coal and **graphite**, has been known for many centuries. However, it was not realized until 1779 that these three substances were composed of essentially the same element. This discovery was made by Carl Scheele, who proved that the mineral graphite was a form of carbon. Diamond is another form of carbon, with a more closely packed atomic structure, which has been subjected to extreme heat and pressure. A more recent artificial form of carbon, **fullerene**, consists of ultra-thin sheets of graphite rolled into cylinders; these are the 'carbon fibres' that are now widely deployed in industry because of their great strength – they are about 100 times stronger than steel.

Carbon is the fifteenth most abundant element; its crustal abundance is difficult to estimate but is of the order of 200ppm (*c.*0.02%). Its atomic structure enables it to combine with other elements in many different ways, and it can form millions of different compounds, especially with oxygen, hydrogen and nitrogen, including the organic hydrocarbon molecules that are the building blocks of life. The two naturally occurring forms of carbon, graphite and diamond, could not be more different – diamond is the hardest substance known, whereas graphite is soft enough to draw or write on paper. Carbon occurs widely in the form of carbonate rocks and carbon-rich sediments like coal, but is more conveniently sourced from graphite deposits such as those in China, India and Brazil (table 7.1H). Over 1 million tonnes were produced in 2012.

The radioactive isotope, carbon-14 (^{14}C), with a half-life of 5730 years, occurs naturally in proportions of one part per trillion, and is the basis for the $^{12}C/^{14}C$ ratio used widely in archaeological dating.

Carbon has numerous industrial applications including electrodes, pigments, and absorbent filters in gas masks. Iron, tungsten and silicon carbides are employed in abrasives because of their extreme hardness. There are large numbers of useful organic compounds based on carbon: for example, hydrocarbons are required for refrigerators, lubricants and solvents; plastics are produced from petrochemicals; sugars and alcohols are based on combinations of carbon with oxygen and hydrogen, and amino-acids on combinations with sulphur.

Diamonds are exploited industrially as abrasives, although a significant proportion of these are made artificially. They are found naturally in the volcanic rocks known as **kimberlites**, which are formed at great depth under extremely high pressure and temperature. Although they were initially sourced from placer deposits, almost all of the present-day supply comes from kimberlite bodies, which occur mainly in southern Africa, Siberia and Australia. Diamond production is measured in 'carats', which are equivalent to 200 milligrams. 75 million carats were produced in 2012, mostly for industrial use. Top four producers were Botswana, Democratic Republic of Congo and Russia (table 7.1I). At an average price of $99 per carat, the diamond trade is obviously very lucrative; when cut for jewellery, the value of the better-quality diamonds increases greatly.

Carbon occurs naturally in the atmosphere in the form of carbon dioxide (c.400ppm.), and is absorbed by plants, which eventually release it into the soil, or into the atmosphere after the plants are eaten by animals or burnt; part is buried to become coal or petroleum deposits. Carbon dioxide is also dissolved in the oceans, where it makes up between 64 and 107ppm (see chapter 10). Some of this carbon is absorbed by marine animals to make calcium carbonate, which is deposited in carbonate rocks. This relatively stable 'carbon cycle' has been thrown out of balance by human intervention: the burning of hydrocarbons in the form of various kinds of fossil fuel (oil, natural gas, coal, etc.) has led to an increase (almost a doubling) of carbon dioxide in the atmosphere, which has been widely blamed for 'global warming'. This subject is discussed in more detail in chapter 12.

Nitrogen (N) (7) makes up 78% of the Earth's atmosphere due to the emission of volcanic gases early in its history. Despite its abundance, it was not discovered until 1772, when Daniel Rutherford realized that what was left in a container of air after all the oxygen had been removed was a separate element. Nitrogen is useful because it is relatively inert and can prevent damage caused by exposure to oxygen. Thus fruit can be stored in nitrogen for up to two years without showing any signs of decay. It also acts as a refrigerant (it boils at minus 196°C) and for this reason is employed in the transport of biological samples.

The human body contains 3% of nitrogen in various proteins and amino-acids including DNA, and nitrogen in the form of nitrates is also an essential nutrient for plants, which extract it from dead organisms in the soil via their roots. The addition of nitrogen to exhausted soil from ammonium nitrate fertilizer is essential in enriching land whose natural nitrogen content has been depleted by agricultural overuse. To counterbalance this benign use of nitrogen is its role in high explosives; nitrogen compounds release their nitrogen easily into the air, hence nitroglycerine, made by combining glycerine with nitric acid, is an explosive liquid that can be detonated merely by impact, and is the basis for dynamite, famously invented by Alfred Nobel.

Despite its abundance in the atmosphere, nitrogen makes up only around 20ppm of the crust, mostly in the form of sodium or potassium nitrates. These were originally mined, but nitrogen is now produced either by fractional distillation of air, or as a by-product of oxygen production, but is marketed in the form of **ammonia**, made by combining nitrogen and hydrogen under high temperature and pressure. Most production is destined for fertilizers. World production in 2012 was 140 million tonnes, the top four producers being China, India, Russia and the USA (table 7.2A).

Oxygen (O) (8) is the third most abundant element in the Universe and is the most abundant in the Earth's crust, making up nearly 50% of its mass, including 86% of the oceans, and 21% of the atmosphere. Three scientists were involved in the discovery of oxygen: Carl Scheele, Joseph Priestly and Antoine Lavoisier, but it was Lavoisier in 1777 who realized that it was a separate element and named it oxygène.

Oxygen is highly reactive and combines with most other elements to form oxides. It is essential for the

country	total kilotonnes
China	45,200
India	12,000
Russia	10,000
USA	8,730
Trinidad & Tobago	5,250
Indonesia	5,100
Ukraine	4,200

A. Nitrogen

country	total kilotonnes
China	95,300
USA	30,100
Morocco	28,000
Russia	11,200
Brazil	6,750
Jordan	6,380
Egypt	6,240

B. Phosphorus

country	total kilotonnes
China	9,900
USA	9,000
Russia	7,270
Canada	5,910
Saudi Arabia	4,090

C. Sulphur

country	total kilotonnes
Jordan	200*
Israel	174
China	100
USA	W

F. Bromine

country	total tonnes
Japan	755
Germany	650
Belgium	200
Russia	145
Canada	144

D. Selenium

country	total kilotonnes
China	4,400
Mexico	1,200
Mongolia	471
South Africa	225

E. Fluorine

country	total kilotonnes
USA	133
Algeria	15
Qatar	13

H. Helium

country	total kilotonnes
Kazakhstan	21.317
Canada	8.999
Australia	6.991
Niger	4.667
Namibia	4.495
Uzbekistan	3.0

I. Uranium

country	total kilotonnes
Chile	17.5
Japan	9.3

G. Iodine

Table 7.2 Production of non-metals by main producing countries – II
Data sources: A-H, US Geological Survey (USGS), 2014; I, World Nuclear Association, 2013. W, data withheld.

as the burning of fossil fuels, have the potential to disturb this balance but may not have done so as yet.

Human toleration of oxygen is contained within fairly narrow limits: below a certain proportion, as high-altitude climbers know, either the oxygen level has to be topped up, or the body has to gradually acclimatize to the more rarefied air; however, more than about 50% is potentially harmful. **Ozone**, the molecular form of oxygen in which three atoms are combined instead of two, is actually toxic; however, it is useful as an anti-bacterial disinfectant. Ozone forms a layer in the stratosphere (see Fig. 2.4B) where it performs the valuable role of shielding the Earth's surface from potentially harmful ultraviolet radiation.

There are two naturally occurring isotopes of oxygen, ^{16}O and ^{18}O, and because the heavier isotope ^{18}O evaporates more slowly, it is relatively concentrated in colder waters, so that the $^{18}O/^{16}O$ ratio in the carbonate shells of fossil marine organisms can be used to track temperature changes in the geological record.

Oxygen occurs in a wide variety of minerals, including the ubiquitous silicates, but for industrial purposes it is produced mainly by the fractional distillation of liquid air, or by passing air through a zeolite sieve, which absorbs nitrogen. More than half is used for steel production, but there are a wide variety of other applications including plastics manufacture, welding and cutting equipment, and rocket fuel. About 100 million tonnes are produced annually, from the principal industrial countries. Only the cost of energy restricts the supply.

metabolic processes of animal cells. However, in geological terms, this role is comparatively recent. Early life on Earth for about the first 2000Ma was in the form of bacteria, which extracted carbon dioxide from the early atmosphere as part of the photo-synthetic process, and excreted oxygen as a by-product. Not until about 500Ma ago did the proportion of oxygen in the atmosphere approach its present levels, enabling the subsequent evolutionary explosion of life forms to take place. Since then, the amount of oxygen in the atmosphere has varied between *c.*15% and *c.*35%, and there is a balance between the amount produced by photosynthesis and the amount used up by respiration and decay. Certain human activities, such

Phosphorus (P) (15) has a bad reputation, and is often referred to as 'the Devil's element'. The uncombined element, in the form of 'white phosphorus', is highly toxic and will cause death within days if ingested; it is the main component of nerve gas. In the form of a phosphate, however, it is essential to the human body: bones and teeth are made of calcium phosphate, and other phosphate molecules in the form of adenosine triphosphate (ATP) play a vital role in storing energy in cells. Excess phosphorus is excreted as urea, and urine was the original source of phosphorus when first discovered. It is also essential to plants, and phosphate fertilizers are required to replenish depleted supplies of phosphorus in the soil.

Although phosphates had been known for centuries, the discovery of the element is credited to Hennig Brand in 1669, although its significance as a new element was only recognized in 1777 by Antoine Lavoisier; it was named after the Greek *phosphoros* meaning 'light-bearer'. Because white phosphorus catches fire spontaneously in air, it was used in match heads, and more recently in tracer bullets, incendiary bombs and napalm. The city of Hamburg was destroyed by the Allies in the Second World War by burning phosphorus. Other industrial applications include detergents (as sodium phosphate), pesticides (as chlorides and sulphides) and, alloyed with copper, in **phosphor bronze**.

The crustal abundance of phosphorus is about 0.1%, and it occurs naturally in the form of phosphates of aluminium and calcium such as **apatite** (hydrated calcium-fluorine-chlorine phosphate) and **monazite** (cerium-lanthanum-yttrium phosphate) which are concentrated in granite pegmatites and alkaline igneous rocks. Initially the main sources of phosphates were deposits of bird and bat **guano** on tropical islands, and some of these are still mined; but natural apatite is the main source now. World production of phosphate in 2012 was estimated at 217 million tonnes, mostly destined for fertilizers. The top seven producers were China, USA, Morocco, Russia, Brazil, Jordan and Egypt (table 7.2B). The US Geological Survey estimated reserves to be about 67 billion tonnes in 2012, and concern has been expressed about possible depletion of this resource. However, 'urine diversion' is beginning to be seen as an environmentally-friendly alternative source, where phosphorus is separated from waste water, which contains significant proportions.

Sulphur (S) (16) occurs naturally in the native form as bright yellow crystals (Fig. 7.2A, B) and has been known and exploited since ancient times as a fumigant and a balm for skin infections. The early name for sulphur was 'brimstone' – mentioned in the Bible in connection with the destruction of Sodom and Gomorrah, on which 'fire and brimstone rained down'. The Chinese used it in their discovery of gunpowder. However, it was Antoine Lavoisier in 1777 who first proved that sulphur was an element.

Despite the attractiveness of native sulphur, its compounds hydrogen sulphide and sulphur dioxide have a bad reputation. Hydrogen sulphide is the source of many bad smells! The unpleasant 'rotten eggs' odour is caused by the bacterial breakdown of organic matter; some bacteria ingest sulphates and excrete hydrogen sulphide. Sulphur dioxide produced by the burning of fossil fuel is the cause of acid rain – actually dilute sulphuric acid – which acidifies groundwater with potentially damaging effects on the ecology of freshwater lakes and rivers.

Sulphur occurs naturally in deposits around hot springs and active volcanoes (Fig. 7.2A), and it was these sources that were exploited historically – Sicily was once an important supplier. Sulphur also occurs in a large number of compounds: the sulphides of metals such as iron, copper and lead, for example, and as sulphates in evaporite deposits. Crustal abundance has been estimated at between 350 and 520ppm. Sulphur is an essential component of amino-acids and proteins and is also a major nutrient in plant growth.

In modern times the element is generally recovered by the removal of hydrogen sulphide from natural gas; it is then usually converted into sulphuric acid. The numerous industrial applications include fertilizer (as calcium sulphate), vulcanization of rubber, antibacterial drugs (e.g. the sulfonamides), and preservatives (as sulphites). World consumption of sulphur in 2012 was estimated at about 68 million tonnes, the top producers being China, the USA, Russia and Canada (table 7.2C).

Selenium (Se) (34) was discovered in 1817 by Jöns Berzelius and Johan Gahn as a by-product of sulphuric acid production. It is an essential element for the human body but is toxic if ingested to excess. It forms part of an anti-oxidant enzyme and is also important in hormone production. Selenium deficiency has been

Figure 7.2
A Sulphur deposits at a fumarole, Vulcano, Lipari, Sicily. Image credit: © Shutterstock, Silky.
B Native sulphur crystals. Image credit: © Shutterstock, Nathalie Speliers Uferman.

shown to lead to an increased risk of cancer and to low sperm production. It occurs in variable quantities in soils, and both low and high concentrations can cause problems for farming. High concentrations in certain plants are toxic to grazing animals.

Crustal abundance of selenium is about 0.05ppm. It occurs mainly in metallic sulphide ores, where it replaces sulphur, and is usually produced as a by-product of copper refining. The principal commercial applications are in glass-making, pigment manufacture, and in photo-electric cells, because its electrical resistance is dependent on the amount of light. Formerly widely used in the electronics industry as a semi-conductor, it has now largely been replaced by silicon. Global production in 2011 was estimated at 2 kilotonnes; 2012 production was mainly from Japan, Germany and Belgium (table 7.2D). The reserves are dependent on the extent of metallic sulphide mining (see chapter 5).

The halogens

This group of non-metals consists of the well-known elements fluorine, chlorine, bromine, and iodine, together with the rare astatine. Both fluorine and chlorine are gases at standard temperature, whereas bromine and iodine are liquids. The unstable astatine is highly radioactive.

Fluorine (F) (9) is one of the most dangerous elements, reacting with almost all the others – explosively, in the case of hydrogen and some carbon compounds. Just 0.1% in air is lethal. It was noted in the sixteenth century that fluorspar (or fluorite – calcium fluoride) would melt and flow when heated,

which led to its use as a flux in metal working. Several chemists died in the attempt to isolate fluorine before Henri Moissan succeeded in obtaining it by low-temperature electrolysis in 1886, for which he was eventually awarded the Nobel Prize.

Although not an essential element for animals, fluorine in the form of a fluoride has been added to toothpaste and, more controversially, to drinking water, for decades in order to prevent tooth decay. The fluoride forms a coating of hard **fluor-apatite,** which protects the teeth from bacterial attack. However, only small amounts (less than about 3mg per day) can be tolerated, and its overuse by young children can cause bone deformity.

The main industrial applications of fluorine are: as hydrogen fluoride, in the manufacture of hydrocarbons for refrigerants; specialized plastics such as Teflon™; breathable clothing materials; in metal smelting, and for uranium enrichment in the nuclear industry. Its role in organic fluorides has decreased dramatically since it was found that **CFCs** (chloro-fluoro-carbons) damaged the ozone layer, with potentially serious effects on the Earth's protection from ultraviolet radiation. CFCs have been strictly regulated since 1987 and have been largely replaced by **HCFCs** (hydro-chloro-fluoro-carbons) which themselves are to be replaced in time by **HFCs** (hydro-fluoro-carbons). The fluoro-carbons are also significant greenhouse gases (see chapter 12).

Fluorine is comparatively abundant, making up around 600ppm of the crust. The main source is fluorite (Fig. 7.3), which occurs in late-stage granitic pegmatites and in hydrothermal deposits,

Figure 7.3 Fluorite (fluorspar) crystal
Image credit: © Shutterstock, Madlen.

often associated with carbonates, along with lead, zinc, silver and barium ores. Fluorine also occurs as fluor-apatite (fluorine-calcium-phosphate) and **cryolite** (sodium-aluminium-fluoride). Fluorite deposits are mined in many countries, the top three suppliers being China, Mexico and Mongolia (table 7.2E). Total world production in 2012 was over 7 million tonnes with reserves estimated at 240 million tonnes.

Chlorine (Cl) (17) first achieved widespread notoriety during the First World War when the German Army released it as a gas over the Allied trenches; 5000 men are said to have died and 15,000 were disabled as a result. To offset this, chlorine probably saved

many more lives during the intervening century because of its use as a disinfectant. It was first produced by Carl Scheele in 1774 by heating hydrochloric acid with manganese dioxide. The resulting greenish-yellow gas attacked most metals and dissolved easily in water to form the weak acid. However, it was Sir Humphry Davy who recognized it as an element in 1810, giving it its name from the Greek *chloros* for yellow-green.

Bleach made from chlorine was first used in 1897 to disinfect drinking water during a typhoid epidemic, and is still widely employed as a disinfectant today, as the characteristic smell of the swimming baths will testify. In water treatment, chlorine is now applied in the form of sodium hypochlorite. Other industrial applications include the manufacture of plastics such as **PVCs** (polyvinyl chloride) and various solvents. The damaging effects of chloro-fluoro-carbon compounds were referred to in the previous section on fluorine.

Chlorine is a necessary component of all life forms but, if ingested in the uncombined state, will combine with the hydrogen contained by enzymes, causing them to disintegrate, which is the reason for its effectiveness as an anti-bacterial agent. It is toxic to humans above about 30ppm and fatal at 1000ppm.

Chlorine makes up about 145ppm of the crust and about 1.9% of the oceans. It combines with almost all the elements to form chlorides. Sodium chloride (common salt) is the commonest chlorine compound and is used in the manufacture of chlorine by electrolysis of brine from salt deposits. Production of chlorine takes place in all the main industrial countries – total global production in 2006 was 65 million tonnes; the USA alone then made 13.8 million tonnes. Production is dependent on salt resources, discussed in chapter 8, which can be regarded as effectively unlimited.

Bromine (Br) (35) is a reddish-brown, fuming liquid – one of only two elements, the other being mercury, that are liquid at room temperature. Like chlorine, it is highly toxic and reacts readily with metals, although not quite as strongly as chlorine. It is also very soluble in water and is present in significant amounts in sea water. It was discovered independently by Antoine-Jerome Balard and Carl Lövig in 1825–26, the former being credited with the discovery. Balard produced the bromine liquid by bubbling chlorine gas through a solution of seaweed ash, and the element is still produced from brine by essentially the same method. The name comes from the Greek *bromos* for 'stench'.

Various bromine compounds have been exploited for centuries; the Roman dye known as 'imperial purple' was obtained from the mollusc *Murex*, which contains bromine. In the nineteenth and early twentieth centuries, bromide salts (both sodium and potassium bromide) were employed as sedatives. More recently, bromine has been used in organic bromo-carbon compounds in fire extinguishers and flame-retardant materials, and methyl bromide as a soil fumigant. However, these bromine compounds are being phased out because of their role in atmospheric ozone depletion. Bromine is necessary for tissue development in animals, and is extracted by algae and seaweeds from sea water.

Crustal abundance of bromine is only 2–3 ppm, but it makes up 65ppm of sea water. It is quite rare in rocks, although it occurs as a bromide in silver ores. Commercial production is from halide salts in evaporite deposits such as those of the Dead Sea in Israel and Jordan, and China and the USA are also major producers (table 7.2F). Total world production in 2012 was 505 kilotonnes; reserves are large.

Iodine (I) (53) is a bluish-black solid at room temperature, which readily sublimates to form a purple vapour. It is familiar to many as the orange 'tincture of iodine' used to disinfect open wounds. It was discovered by Bernard Courtois in 1811 but not identified as an element until 1813 – both Gay-Lussac and Humphry Davy claiming the credit.

Iodine is an essential trace element in many animals, being a key constituent of the thyroid gland, which controls several metabolic functions. Iodine deficiency leads to neck swelling (goitre). Although the crustal abundance of iodine is only about 0.5ppm, it is found in significant concentrations in sea water, being extracted by brown algae, the seaweed Laminaria, and other marine organisms. Because iodine is so easily soluble, soils can readily become depleted in it, and peoples without ready access to seafood are therefore vulnerable to iodine deficiency – a condition experienced by up to two billion people worldwide, leading to various disabilities, including intellectual impairment. In developed countries, sodium iodide is often included in table salt, and iodine compounds are also added to animal feed.

The radioactive isotope iodine-131 was released during the Chernobyl nuclear explosion and passed into the food chain via grazing animals. It is thought that up to 4000 people developed thyroid cancer as a result.

Iodine is used industrially as an animal feed supplement, in printing inks and dyes, and in medical procedures as an X-ray contrast material. It is produced either from **caliche** (a natural deposit containing sodium iodate and iodide) found mainly in Chile, or from brines derived from evaporite deposits. About 28 kilotonnes were produced in 2012, mainly from Chile, and Japan (table 7.2G). Chilean reserves are very large but sea water and seaweeds are potentially inexhaustible.

Astatine (At) (85), the heaviest element in the halogen group, is one of the rarest elements on Earth. It is produced by the radio-active decay of uranium and thorium and is so highly radioactive that only about 28g are present on Earth at any one time; its longest-living isotope has a half-life of only 8 hours. It was first produced in 1940 by bombarding bismuth with alpha particles in a cyclotron. The isotope astatine-211 is used in radio-therapy for some cancer treatments.

The noble gases

The noble gases are a group of six non-metals occupying the right-hand side of the Periodic Table (Fig. 2.2). They are all gases at room temperature and are colourless, odourless and non-reactive.

Helium (He) (2) is the second most abundant element in the Universe (after hydrogen), being created during the 'big bang' and in stars such as the Sun as a result of the nuclear fusion of hydrogen atoms (*see* chapter 2). It was origin-ally detected in the Sun's spectrum by Norman Lockyer and Edward Frankland in 1868, and named after the Greek Sun god *Helios*.

Although it makes up 5.2ppm of the atmosphere, its crustal abundance is very low – about 0.008ppm. The main source is natural gas, in which the helium is produced by the radioactive breakdown of uranium and thorium in uranium ores such as pitchblende, from where it escapes and accumulates in natural gas reservoirs.

Helium is used in liquid form as a super-coolant (it is liquid at -269°C), for example, in the cooling of supercon-ducting magnets; it is also used, because of its low density and non-reactivity, in balloons and airships, and it protects combustible fuel from accidental igni-tion in spacecraft launching gantries.

Total world production of helium in 2012 was 174 million cubic metres; mostly in the USA (table 7.2H), where there are large reservoirs. However, concern has been expressed that these reserves could become exhausted by 2018, and that it may become neces-sary to restrict its use in non-essential applications. Helium is a finite resource and once released, eventually escapes into space. Nevertheless, the other producing countries do have substan-tial reserves and these may well last at least until the end of the century.

Neon (Ne) (10) for most people is associated with neon lighting, which has transformed the centres of our cities with brightly-coloured advertis-ing displays. Although a colourless gas, like the other noble gases, neon emits a bright orange-red glow when an electric current is passed through it. It is totally non-reactive, and com-bines with none of the other elements. Consequently it has no biological role.

Neon was discovered in 1898 by Sir William Ramsay and Morris Travers by the fractional distillation of liquid air. Once the more volatile constituents – nitrogen, oxygen and argon – had been removed, only the rare gases, neon, krypton and xenon, remained and were identified as separate ele-ments by their spectra. The name is derived from the Greek *neos* for 'new'.

Apart from lighting, neon is used in lasers and high-voltage switching gear. It is also a highly efficient refrigerant, liquefying at -246°C.

Neon is fifth in cosmic abundance but is rare on Earth, comprising only 18.2ppm of air by volume. It has been detected in volcanic gases and thermal springs, but because of its inability to form compounds and its volatility, it readily escapes into the atmosphere. It is extracted commercially by the fractional distillation of liquefied air. Only about 375 million litres of neon was produced in 2007, about 70% from Russia and other members of the former Soviet Union. Only the expense of extraction limits future supply.

Argon (Ar) (18) was discovered in 1894 by Lord Rayleigh and Sir William Ramsay by extracting it from liquid air. It was named after the Greek *argos*, meaning 'inactive' or 'idle' in reference to its lack of reactivity with other elements. Like neon, it is used in illuminated signs, giving an electric blue light when a current is passed through. Conventional elec-tric light bulbs also use argon to provide an inert atmosphere. For the same reason, it is used in welding to prevent oxidation, and also in steel smelting, where it is blown through the molten metal along with oxygen; while the oxygen combines with carbon, the argon minimizes oxidation of the more valuable trace elements such as chromium. Other applica-tions include insulation in double glazing, and as argon lasers in surgical procedures to weld arteries, destroy cancers and correct eye defects.

Argon makes up about 0.93% of air. Most atmospheric argon is produced by the radioactive breakdown of potas-sium-40 in granitic rocks, giving rise to the well-known geological method of potassium-argon dating. Potassium-40 has a half-life of 1250Ma, which makes the method suitable for dating

Precambrian rocks. Argon is 500 times more abundant in air than neon, and is therefore cheaper to produce; annual production in 2007 was 700 kilotonnes. Many of the applications listed above for neon can equally well apply to argon. As in the case of neon, only the expense of the extraction process limits the use of argon.

Krypton (Kr) (36) was extracted from liquid air by Sir William Ramsay and Morris Travers in 1898. Having evaporated nitrogen and oxygen, they proceeded to liquefy the argon gas that remained, and when 15 litres of this had been evaporated, 25 millilitres of liquid gas remained, which they named krypton after the Greek word *kryptos* for 'hidden'. Krypton is much rarer than argon and neon, making up only 1ppm of air. Because it is also heavier than these gases, all the krypton present when the Earth was formed is still here.

Krypton is used in the same way as neon and argon in lighting, giving out a brilliant bluish light when an electric current is passed through it. It is used in photographic flash lights and in strobe lighting because of its rapid response to an electric current. The krypton–fluoride laser gives out a deep violet light. Krypton has also been found effective in energy-saving light bulbs because it reduces the rate of evaporation of the tungsten filament, enabling it to burn at a lower temperature for a longer period, thus producing more light than heat.

The radio-active isotope krypton-85 is produced in nuclear reactors and escapes into the atmosphere, giving rise to a method of monitoring nuclear activities.

Crustal abundance of krypton is about 1ppm, almost all in the atmosphere. Eight tonnes were produced in 2007. Again, only extraction costs limit its exploitation.

Xenon (Xe) (54) was the last of the noble gases to be discovered, in 1898 by William Ramsay and Morris Travers. After krypton had been removed from their liquid air sample, an even heavier gas remained, which they named xenon, from the Greek *xenos* for 'stranger'.

Xenon was thought initially to be inert until it was found to combine with fluorine and platinum. Its use is limited by its rarity (crustal abundance is only 0.08ppm) and the expense of extraction, but it is employed in the same way as krypton in gas discharge lamps and in high-speed photography. It is considered to be ideal for the 'ion engine' used to power satellites and deep space probes. This device works by accelerating ionized xenon gas and expelling it at extremely high velocities (up to around 40,000km/sec) to produce the thrust. Only about 0.6 tonnes were produced in 2007.

Radon (Rn) (86) was first discovered by Friedrich Dorn in 1900 when he noted that radium compounds emitted a radioactive gas. Two years later, Ernest Rutherford and Frederick Soddy studied the radioactive gas emitted by thorium and declared it to be a new element, the heaviest of the noble gases. Radon gas is produced by the radioactive breakdown of uranium and thorium in granitic rocks and diffuses upwards into the atmosphere. If it collects in a confined space, it can be dangerous – contributing, it is thought, to a small proportion of lung cancers. Nevertheless, in the early twentieth century many people thought that to drink radon water (radon gas dissolved in water) was a health benefit!

In the 1990s the UK government issued home detector kits to check for dangerous levels of radon-induced radiation. The longest-living isotope of radon gas, radon-222, has a half-life of 3.8 days and decays into solid radioactive particles that themselves emit harmful alpha radiation. There are only a few tens of grams in the atmosphere at any one time. Homes built from, or on, granite suffer the most risk from radon if not well ventilated, and it is particularly dangerous in deep uranium mines. Radon is measured in **becquerels** per cubic metre (1 becquerel = 0.7×10^{-15} grams), and the maximum acceptable level in homes was recently declared by the UK Government to be 200 becquerels per cubic metre. Radon has been used to a limited extent in radiation therapy but has largely been superseded.

The actinides and the trans-fermium elements

The elements numbered 89–103 in the Periodic Table are known as the **actinides**, after actinium, number 89 (see Fig. 2.2). All are highly radioactive and because natural reserves of most of them have decayed completely, they can only be produced in the laboratory. The only elements in this group that have significant industrial applications or resource implications are actinium, thorium, uranium, plutonium and californium. The elements numbered 101–118 (including the last three actinides) are known as the 'trans-fermium elements', none of which occurs naturally; these can only be manufactured (expensively) in the laboratory. They appear to be of academic interest only and do not have any significant applications.

Actinium (Ac) (89) makes up 0.2 parts per billion of uranium ore. Its longest-lived isotope, actinium-227, has a half-life of 21.8 years. Despite the high cost of production, it has found an application as a neutron source for laboratory experiments and in targeted radiotherapy.

Thorium (Th) (90) is the most abundant of the actinides, with a crustal abundance of 9.6ppm – three times more abundant than uranium – and occurs in uranium ores. It was discovered in 1829 by Jöns Berzelius, and its radioactivity was subsequently investigated by Marie Curie. Thorium-232 has a half-life of 14 billion years, which means that most of the primordial supply of the element still exists. The alloy Mag-Thor with magnesium is strong, light and resistant to high temperatures, and is used in missiles and aircraft construction. It is considered to be a potential fuel source for nuclear fission reactors as a safer alternative to uranium because it does not produce long-lived radioactivity. There is much less risk of accidental meltdown with a thorium reactor, and there is the additional advantage that, because it requires small amounts of the dangerously radioactive plutonium, it provides a convenient method of disposing of the World's waste plutonium stocks.

Uranium (U) (92) is the best-known of all the radioactive elements because of its use in the atomic bomb, and in the nuclear power industry. Uranium oxide was discovered in 1789 by Martin Klaproth, who extracted it from pitchblende, but it was not until 1841 that it was isolated as an element, by Eugène-Melchior Peligot; its radioactivity was discovered in 1896 by Antoine Becquerel, who noticed that a photographic plate left in a drawer along with a sample of uranium had become fogged.

There are several radioactive isotopes of uranium, two of which occur naturally: uranium-238 and uranium-235. Uranium-238 is the more abundant and has a half-life of 4.5 billion years, making it suitable for radioactive dating of very old rocks, and even of the age of the Earth itself. Uranium-235, which makes up only 0.7% of natural uranium, is the isotope used in nuclear fission, which means that it has to be expensively separated from the more abundant isotope. Uranium is a major source of the Earth's internal heat, driving plate tectonics and vulcanicity. Early uses of uranium, before its radioactivity and toxicity had been established, were in pigments for pottery and glass, but more recently the main applications have been in nuclear weapons and nuclear power. One kilogram of uranium is equivalent to 13 kilotonnes of conventional explosive, and to 1500 tonnes of coal when used as a fuel. Depleted uranium (i.e. after U-235 has been extracted) is used in armour-piercing projectiles because of its weight.

Uranium is an incompatible element, present in highly-evolved granitic melts. It is easily soluble, and accumulates in sedimentary deposits, especially at unconformities above granitic basement (see chapter 4); the principal primary ore is pitchblende or uraninite (uranium oxide). The crustal abundance of uranium is about 3ppm, and around 58 kilotonnes were mined in 2012; the top three producers were Kazakhstan, Canada and Australia (table 7.2I). There are large reserves.

Plutonium (Pu) (94) is well known (or notorious) because of its role in the atomic bomb. Just 6.2 kilograms were enough to destroy Nagasaki in 1946. It was discovered in 1940 at Berkeley, USA, by bombarding uranium-238 with deuterons (heavy hydrogen nuclei) but is not found naturally; all the elements beyond neptunium (93) in the Periodic Table have to be produced artificially in the laboratory. In addition to its military applications, plutonium has been used as a power source in spacecraft.

Californium (Cf) (98) is also produced artificially; its isotope, californium-252, has an important application in the technique of neutron-activation analysis, employed in the search for precious metals and oil, and to scan baggage for explosives.

Over the many millennia that the human race has inhabited our planet, a use has been found for almost everything that is to be found on it. Previous chapters have detailed the different types of mineral resources based on individual elements. However, the rocks themselves, which may contain many different minerals, are also a resource, and are used for different purposes and in different ways. One of the earliest uses of natural rock materials was for building shelters, and the exploitation of various types of rock directly for building purposes is still widespread, particularly in less well-developed countries. Some types of rock are employed indirectly in the construction industry; these include limestone (to make cement), clay (for brick-making), and sand and gravel (for concrete-making). Other types of rock used industrially include rock salt and gypsum from evaporite deposits. Rock types exploited for fuel, (e.g. coal, oil, etc.) are discussed in chapter 9.

Building stone

The first peoples to settle in a particular area would have used the most suitable building material that was readily available, and in many parts of the world this is still the case. There is a wide variety of such material; for example, the beautifully constructed dwellings of the Skara Brae settlement in Orkney (Fig. 8.1A), dating from the fourth millennium BCE, may be contrasted with the crudely-built houses made from turf and rounded field stone used locally until the nineteenth century in other parts of northern Scotland. The difference is due mainly to the quality of the material available locally: thin slabs of Old Red Sandstone in Orkney, compared with rounded boulders of gneiss and plentiful supplies of turf in Ross-shire and Sutherland.

Almost any type of rock may still be used for building purposes, although in the more industrialized countries, bricks and concrete have replaced stone to a large extent. Rock suitable for good building stone is not as common as might be thought. Such a building stone needs to be hard and have high mechanical strength. It should ideally also be mineralogically pure, free from undesirable alteration products, and from minerals that would be susceptible to weathering, or to atmospheric pollution. A good building stone must be easy to cut or split and shape, and the spacing of joints and bedding planes in the source-rock formation is therefore important in its selection.

Building stone may be used either as the main construction medium, as is the case of many of the older grand public buildings in most cities, or more commonly now as **cladding**, where thin slabs of stone are used as a covering to hide or beautify a more ordinary building material.

Igneous rocks have always been widely used as building material because of their hardness and resistance to weathering, but suffer the disadvantage of being difficult (and expensive) to work. Most types of igneous rock are suitable, provided that they are completely crystalline and consolidated, such that the constituent minerals are bound together by a crystalline matrix. Most types of igneous rock, including basalt and gabbro, are generally referred to as 'granite' by the building trade. Igneous rocks are especially favoured for monuments and gravestones because of their attractiveness and durability. Figure 8.1B illustrates the magnificent second century BCE Egyptian granite Sphinx at St Petersburg, Russia.

China is by far the biggest exporter of 'granite', followed by India, Indonesia and Italy (table 8.1A).

Limestones (including **dolostones**) are also widely used as building stone because of their relative ease of working; both calcite and dolomite, the main constituents of limestone, are softer, and easier to cut, than igneous rocks, but are susceptible to chemical weathering; sulphur dioxide in the atmosphere forms sulphuric acid, which will dissolve calcite. Despite this disadvantage, many of the most beautiful buildings ever constructed have been built with limestone: for example, the Parthenon in Athens (Fig. 8.1C).

The term 'marble' means a metamorphosed limestone to a geologist,

Figure 8.1 Rock as a building material

A Skara Brae, fourth century BCE settlement, Orkney, built with local sandstone. Image credit: © Shutterstock, duchy. **B** 3500-year-old Egyptian granite sphinx, transported to St. Petersburg by Catherine the Great. Image credit: © Shutterstock, Eugene Sergeev. **C** The Parthenon, Acropolis, Athens, built in 438BCE from the marble quarries of Mount Pentelikon. Image credit: © Shutterstock, S. Borisov. **D** Victorian red sandstone tenements, Edinburgh. Image credit: © Shutterstock, Drimafilm.

country	million tonnes
China	6,500
India	2,500
Indonesia	1,500
Italy	1,000

A. granite

country	million tonnes
Turkey	2,500
Italy	1,500
Spain	1,250
Croatia	1,125

B. limestone

country	million tonnes
Spain	650
China	250
Kyrgystan	200
Brazil	200
India	150

C. roofing slate

country	million tonnes
China	2,210
India	270
USA	74.9
Iran	70
Brazil	68.8
Turkey	63.9

D. cement

country	kilo-tonnes
USA	4,980
Greece	800
Brazil	567
Turkey	400
Germany	375

E. bentonite

country	kilo-tonnes
China	70,000
USA	37,200
India	17,000
Germany	11,900
Australia	10,800
Canada	10,800
Mexico	10,800

F. salt

country	kilo-tonnes
China	48,000
USA	15,800
Iran	13,000
Thailand	9,000
Spain	7,100

G. gypsum

country	kilo-tonnes
USA	22,871
China	19,639
Russia	4,778
Canada	4,470
Japan	4,369

H. bitumen

Table 8.1 Production of building materials

A–C, export data; D–E, production data. Sources: A–C, UN Comtrade database, 2005; D–G, US Geological Survey, 2014; H, UN, 2010.

but is routinely used in the building industry for a good quality limestone that can be polished to a smooth surface. Many such polished 'marbles' contain fossils that add to their attractiveness. Much building limestone is still produced in Italy (including the famous Carrara marble), with Turkey, Spain and Croatia also major exporting countries (table 8.1B).

Sandstones are commonly used as building stone, but are very variable in quality. Some can be porous, and often contain impurities such as iron compounds, which may weather to produce unsightly brown staining. Sandstones whose constituent grains are bound together by a crystalline cement such as calcite or quartz are less susceptible to erosion. The attractive red sandstones of Devonian and Permian age were widely used in Scotland in Victorian and early twentieth century times (Fig. 8.1D). Only India is now a major producer of building sandstone, exporting 500 million tonnes in 2005.

Metamorphic rocks

Slate (metamorphosed mudstone) was routinely used in the past as a roofing material, where available, because of its impermeability and its ability to split along the cleavage planes to produce thin slabs. However, it has largely been replaced in many countries by artificial materials, such as clay or concrete tiles. There is still a large market for slate in Europe, which is mostly sourced from Portugal and Spain. The main exporting countries for roofing slate in 2005 were Spain, China, Kyrgyzstan, Brazil and India (table 8.1C).

Some types of metamorphic rock are employed as ornamental stone, because of their attractive colour or texture. **Serpentine** and coloured marble are both used for this purpose.

Resources

The supply of building stone is effectively infinite; however, sourcing good quality building stone in highly developed, densely populated countries is subject to severe limitations imposed by environmental constraints. Total world production of building stone was estimated at 68 million tonnes in 2001, 65% of it from Europe. The main producers were China, Italy, India, Iran and Spain.

Cement

This material, vital to the construction industry, is made by heating crushed limestone with clay to produce a fine grey powder. This then hardens to a rock-like consistency when water is added. Large amounts of relatively pure limestone are required to produce cement in sufficient quantities, and limestone quarries are a familiar part of the landscape in many parts of the world.

Cement was widely used by the Romans, along with powdered volcanic ash, to make concrete, with which large structures such as aqueducts were built. From the eighteenth century in Europe, lime and clay were mixed to make **hydraulic cement**, the addition of the alumino-silicate minerals in the clay giving the cement the ability to set more quickly in the presence of water. The cement is manufactured by burning crushed limestone at over 800°C, which drives off carbon dioxide, leaving calcium oxide (**lime**). When water is added to lime, it forms calcium hydroxide (**slaked lime**) and the carbon dioxide in the air causes the cement to slowly harden to calcium carbonate again with the consistency and strength of the original rock.

Cement is used in the building industry in three main ways: directly as a covering material (render) to protect brick walls from weathering; mixed with sand to make **mortar**; and with sand and **aggregate** to make concrete. Specialized cements are made for specific purposes: for example, by adding various compounds to increase its strength.

Total world production of cement in 2012 was 3800 million tonnes, the foremost producing countries being China and India (table 8.1D).

Hazards and environmental concerns

Cement powder is hazardous to the skin, eyes and lungs, and must be handled with care. The production process causes environmental problems both at the quarrying stage, from dust, noise and blasting effects, and at the production stage when the high-temperature combustion releases carbon dioxide and various other pollutants into the atmosphere. The nature of the pollutants depends on the materials added to the lime; for example, the clays used may contain heavy metals such as cadmium or mercury. 'Green cement' is now increasingly being manufactured, using various types of waste material to reduce the emissions of carbon dioxide; for example, the addition of sewage sludge to the high-temperature kilns neutralizes the organic pollutants.

Industrial clays

Clay is a type of fine-grained, sedimentary rock consisting mainly of clay minerals; these are hydrated alumino-silicates with a sheet structure similar to mica. Clay has the same range of composition as mudstone and shale, but differs in containing enough moisture to be plastic and workable. There are many industrial applications, including brick and tile making, pottery and ceramics. A good industrial clay should contain a suitable proportion of clay minerals to enable it to be successfully worked and fired; brick clays require a lower proportion of clay minerals than ceramics, for example. The firing process drives off the water and transforms the material to one composed of hard crystalline silicates just like an igneous rock. Clays are also used as a natural seal, in dam construction, for example, because of their impermeability.

Brick and tile clays

In many parts of the less industrialized world, clay or mud from any convenient local source is used for building, and the material merely dried in the sun. However, the clays used in more sophisticated brick and tile manufacture, though often sourced relatively close to where they are needed, are fired in large kilns. The process requires the clay to fall within a certain range of compositions, although it may still contain impurities such as iron compounds.

Pottery and ceramics

Clay has been used to make pottery since prehistoric times, and the design of the pottery is used by archaeologists to date the settlements where it is found. Some of the earliest has been dated at 14,000BCE in Japan. The clays used in pottery are very variable in composition and quality, and this is reflected in the colour and texture of the pottery. The purest type of ceramic clay, known as **china clay**, has long been used in the ceramics industry, and is derived from hydrothermally altered, deeply weathered granites, such as the well-known granites of Cornwall in SW England. This pure clay is composed of the clay mineral **kaolinite**, an alteration product of feldspar, and is extracted from the weathered rock by washing, which separates out the kaolinite from the accompanying quartz and mica.

Bentonite is a specialized type of clay consisting mainly of the clay mineral **montmorillonite**, and is typically formed from the weathering of volcanic ash; there are large deposits of Cretaceous and Cenozoic age in the western USA. Some bentonites contain potassium, sodium or calcium attached to the alumino-silicate structure of the montmorillonite.

Sodium bentonites are used as drilling mud in oil and gas exploration because of the colloidal properties of the clay (i.e. it forms a suspension in water), and because it expands to several times its volume when wet. About half of the bentonite produced in the USA is used for this purpose. Calcium bentonites adsorb fats and oils and are the main ingredient of **fuller's earth**, one of the earliest cleaning agents. Bentonite also forms a bonding medium for sand, which enables it to form moulds for sand-casting purposes. Other specialized applications include wine purification and pet litter.

Bentonite is the most commercially important type of clay. In 2012, total world production was 9950 kilotonnes, of which the USA, Greece, Brazil, Turkey and Germany were the leading suppliers (table 8.1E).

Aggregate, sand and gravel
Aggregate, sand and gravel are used extensively in the construction industry – the largest proportion, together with cement, in the production of **concrete**. The other principal application is in road construction and railway ballast.

Sand and gravel
Most economically viable deposits of sand and gravel are unconsolidated, and may be of fluvio-glacial origin (i.e. deposited from melting Pleistocene ice sheets), or **alluvial**, deposited in modern river valleys and flood plains, or derived from coastal marine deposits. Enormous quantities of this material are produced in every developed country, generally from relatively close to where it is required, because of the costs of transporting it. World production in 2012 was estimated at about 139 million tonnes. The USA was by far the largest producer. Production figures from the UK and USA are given in table 8.2. Resources are very large indeed, and the only major constraint is the impact of extensive quarrying in environmentally sensitive areas.

Sand is exploited in a large number of ways: its recreational value on beaches and in childrens' play-parks is obvious, as is its use in sandbags as a flood defence. However, there are also many industrial applications: mortar is made with sand and cement, and concrete with aggregate and cement. Sand is also used for hydraulic fracturing ('fracking') in oil and gas exploration (*see* chapter 9); to make moulds for molten metal in foundries; added to clay to make bricks; and in sand-blasting. Pure silica sand is required for glass-making.

The composition of sand varies according to the nature of its source. Although the main constituent is usually quartz (silica), it may also contain feldspar, mica or magnetite, for example, and its colour may vary according to the nature of the impurities. Sands from a volcanic source may be quite black, and sands with even small quantities of iron compounds may be yellow, red, or green. Sands in areas where coral reefs are the dominant source rock are composed almost entirely of calcite.

Sand may be sourced either from natural sand deposits, such as beach sand or alluvium, or it can be extracted from sand and gravel deposits. There are often stringent restrictions on the extent to which natural sand deposits can be exploited, and certain countries have banned sand exports because of concerns over the environmental impact of extraction, especially on beaches.

Aggregate
This term is used for rock that is artificially crushed to gravel size for use in concrete making and other construction purposes. Aggregate is preferred to sand and gravel for road building because its physical properties are more predictable. Suitable rock needs to be hard, with a high crushing strength. Limestone and fine-grained igneous rock are widely used for this purpose. Crushed stone is also a by-product of the quarrying of building stone. There is an effectively unlimited supply of such material; however, as in the case of sand and gravel, the principal constraints are environmental, and these apply particularly in highly industrialized and densely populated countries such as the British Isles and other parts of Western Europe.

Rock salt and gypsum
Both rock salt (sodium chloride) and **gypsum** (hydrated calcium sulphate) are extracted from evaporite deposits such as those of Permian age in Western and Central Europe.

Rock salt or **halite** (sodium chloride) is abundant, making up 3.5% of sea water, so that reserves are effectively inexhaustible. Salt is essential for all forms of animal life, which require extra supplies if they are solely reliant on a plant diet. Salt has

been a prized commodity throughout history. There is evidence that it was produced by the Chinese in the sixth century BCE; it is referred to frequently in the Bible; it was widely traded by ship across the Mediterranean by Greeks and Romans, and carried across the Sahara by camel caravans.

Only a small proportion of the salt produced now is destined for human consumption; about 70% is used for various industrial purposes, including the manufacture of caustic soda, chlorine and various plastic products, paper pulp processing, and water conditioning. A considerable amount is employed to keep roads free of ice in winter.

Too much salt is considered to be harmful, being responsible for raising blood pressure and potentially causing strokes and heart attacks. High salt levels in processed foods have caused particular concern.

Salt is produced from the extensive **evaporite** deposits and, in many hotter parts of the world, is recovered from sea water, which is held in shallow tanks and allowed to evaporate. Both unrefined sea salt and evaporites contain chlorides and sulphates of magnesium and calcium as impurities, and these have to be removed by successive stages of evaporation and recrystallization to obtain salt pure enough for human consumption. Total world production of salt in 2012 was estimated at 259 million tonnes, the top producers being China and the USA (table 8.1F).

Gypsum (hydrated calcium sulphate) also occurs in evaporites deposited from sea or lake water, although less abundantly than rock salt. It also forms around volcanic hot springs and in veins within other sedimentary

Figure 8.2 'Desert Rose'
Crystals of gypsum encrusted with desert sand. Image credit: © Shutterstock, jmboix.

rocks. Gypsum may take various forms, including transparent, well-shaped tabular crystals; the variety known as **desert rose** contains sand grains which give it a characteristic texture (Fig. 8.2). **Alabaster** is a massive, fine-grained, white or light-coloured variety, which is easy to carve and has been employed by sculptors since the time of the ancient Egyptians.

Gypsum has many industrial applications, including the manufacture of plaster and plasterboard, splints and moulds for medical purposes, as a fertilizer, in baking, as a dough conditioner, and as a component of cement to improve the setting quality. The plaster used in the building trade is actually composed of **anhydrite**, which is dehydrated, or 'burnt' gypsum.

There are substantial reserves of gypsum in many different countries. Total world production was estimated at 152 million tonnes in 2012, the main producers being China, the USA, Iran, Thailand and Spain (table 8.1G).

Bitumen

Bitumen is the geological name for the substance often termed **asphalt**, **tar**, or **pitch**, and is a sticky, black, viscous or semi-solid form of **petroleum**. Petroleum deposits are formed from buried organic material, and are discussed at length in chapter 9 under fossil fuels; however, bitumen as a resource has quite different properties and industrial applications from other petroleum products. It occurs in natural deposits ('crude bitumen'), such as the famous 'pitch lake' in Trinidad and Tobago, but is also produced artificially ('refined bitumen') as a product of petroleum refining. Almost all the bitumen produced now is refined from petroleum deposits. One of the largest reserves of crude bitumen is the Athabasca bituminous sands in Alberta, Canada, which contain up to 20% of bitumen.

Bitumen has been exploited since the earliest times, mainly for waterproofing purposes. The Bible states that Noah used pitch to make the Ark seaworthy, and many of the early civilizations record its use. It has been deployed as paving material since the 1830s in Britain and France. The most important industrial application now is in road construction, and around 70% of the bitumen produced is mixed with aggregate to form **tarmacadam** ('tarmac') for this purpose. It is also still widely used as a waterproofing medium, as in roofing felt, and to seal joints in other building materials.

World production of refined bitumen in 2007 was 102 million tonnes. The USA and China were the top producers in 2010 (table 8.1H). Reserves of bitumen depend largely on the supply of crude petroleum, which is still plentiful, although there may be restrictions on its exploitation in the future due to environmental concerns. Much of the bitumen used in road surfacing is routinely recycled during re-surfacing or widening operations, so that bitumen resources do not pose a significant problem.

A

aggregates	million tonnes
sand & gravel	868
crushed stone	1,270
total aggregate	2,140

B

aggregates	kilo-tonnes
sand & gravel	55,015
crushed rock	43,826
limestone*	43,032
igneous rock	38,869
sandstone	9,144
secondary**	60,000

Table 8.2 Production of aggregate
A Different types of aggregate produced by the USA in 2010.
B Different types of aggregate produced in the UK in 2011.
Data: A, US Geological Survey; B, British Geological Survey.
Notes: * includes dolomite; ** includes recycled material.

9   Non-renewable energy resources

What is energy?

Energy, work and power

It is important, in considering our energy resources, to explain what energy means to the scientist, since the word has a much broader meaning in everyday usage. **Energy**, strictly defined, means the capacity to perform some **work** – that is, the ability to provide a **force** that will move an object through a given distance, for example. **Power** is the measure of the amount of work that can be done in a given time. Table 9.1 gives definitions of these terms. There are several different forms of energy, the most obvious in the context of the Earth being solar radiation, internal heat and gravitational potential. Many sources of energy are in fact secondary, being derived from these more fundamental forms.

Kinetic and potential energy

All moving bodies possess what is known as **kinetic energy**; this increases with the square of the velocity. Thus a cricket ball travelling at 60 miles an hour has four times the energy of one travelling at 30. Gravitational energy, in contrast, is a type of **potential energy**, where the energy of an object of a given mass is proportional to the height through which it can fall, so that a cricket ball held two metres above the ground has only twice the potential energy of one held one metre above. Hydro-electric

---

ENERGY and POWER

The amount of energy expended by a system is measured by the work done plus the heat generated. Thus:

Energy (E) = **Work** (W) + **Heat** (Q), i.e. E = W + Q

The amount of work done by a system is measured by the product of the force employed on a body of given mass and the displacement of that body. Thus:

Work (W) = **Force** (F) x **Displacement** (s), i.e. W = Fs.

Work is measured in **Joules**, where

1 joule (J) = 1 newton metre (Nm), i.e. J = Nm

Force is the product of a mass (measured in grams) and its acceleration (measured in centimetres per second squared). Force is measured in **Newtons**, where

1 newton = 1 gram cm/second^2, i.e. N = g cm s^{-2}

Heat (thermal energy) is measured in calories (cal), where one calorie = the amount of heat required to raise the temperature of 1 gram of water from 14.5^0 to 15.5^0 centigrade. It is usually recorded in kilocalories (Kcal).

Power is the rate at which work is done. Thus the amount of power produced by a system is the work done (W) divided by the time taken (s). It is measured in watts (w) where 1 watt = 1 joule per second. The capacity of a power system is usually recorded in kilowatts (1000 watts) (Kw) or megawatts (10^6 watts) (Mw) or gigawatts (10^9 watts) (Gw) and the output in kilowatt-hours (Kw s^{-1}) or megawatt-hours.

Table 9.1 Energy and power
Definitions of the terms 'energy' and 'power' as used by scientists and engineers when describing power generating systems such as power stations, etc. SI units are used here although engineers often use alternatives such as pounds, inches, etc.

power exploits the potential energy derived from the height difference between the water in the reservoir and the power station downstream.

Heat

Thermal energy is one of the more obvious forms of energy. Water, for example, when heated to boiling point, provides steam, which expands, creating a force that is able to do work by pushing a piston. Geothermal heat is an important source of energy in volcanically active regions such as Iceland and New Zealand. Thermal energy can also be obtained by exploiting the temperature difference between the warmer conditions at depth and the ground surface.

Electro-magnetic energy

Solar radiation is a type of **electro-magnetic energy**, which is manifested in sunlight but also includes the invisible radiation beyond the visible spectrum. It is by far our largest source of natural energy. The electricity generated in power stations is an obvious form of electromagnetic energy, but is a secondary product of some other energy source.

Chemical energy

This type of energy exploits the **exothermic** nature of certain chemical reactions in which two elements are combined, giving off heat that can then be used as a source of power. For example, the reaction of hydrogen and oxygen to produce water is the basis of the hydrogen fuel unit, which is likely to become a major source of power for future motor vehicles. Much of the chemical energy that is currently deployed is derived from the burning of plant material, either directly, using wood or other biomass, or indirectly, from fossil fuel such as coal. This type of material is an indirect form of energy, which was originally derived through photosynthesis from sunlight. Nuclear energy exploits the chemical energy locked up in the strong bonds that hold the sub-atomic particles together.

Renewable and non-renewable energy

The distinction between renewable and non-renewable energy is at the heart of the current debate about measures to combat climate change. The distinction is not always clear-cut, since some ostensibly renewable materials (e.g. timber) could be depleted if not managed carefully, and wind-generation structures use a significant amount of non-renewable energy in their manufacture. However, other examples are more obvious; fossil fuels are clearly non-renewable, as are most forms of chemical energy. Nuclear energy is an interesting example of a supposedly 'green' energy source (it does not release greenhouse gases) but since it is, at present, reliant on uranium, an exhaustible fuel, it must be considered to be non-renewable.

The principal sources of non-renewable terrestrial energy are: fossil fuels (coal, oil and natural gas) and nuclear processes. Renewable resources are discussed in the following chapter.

Fossil fuels

Fossil fuels are currently the largest suppliers of energy, although their role in global warming has meant that many countries are attempting to reduce their use in favour of more environmentally friendly options. The three main types of fossil fuel – coal, oil and natural gas – are all derived from decayed organic material that has been buried, compressed, subjected to various tectonic processes such as folding and faulting, and now forms part of the geological record. They actually form a spectrum of material, in the sense that oil and gas are merely the more mobile of the products of organic decay, and typically migrate upwards into different rock strata where they may become concentrated and exploited.

Coal

Coal has been exploited as a fuel since at least the Bronze Age in Britain, where there is evidence of its use for funeral pyres, and in the first millennium BCE in China, for smelting copper. The Romans exploited coal in England and Wales, but it was only widely available as a fuel from about the thirteenth century, at first from easily accessible surface deposits, and later mined from shallow shafts. Deep coal mining became common during the Industrial Revolution, and by the 1830s in Britain, coal had replaced watermills as a source of power generation. Deep coal mining in Britain has been replaced by near-surface quarrying ('**opencast mining**') on a greatly reduced scale. Both deep and opencast mining for coal are still carried out extensively in many countries, despite environmental concerns.

Types of coal

Coal embodies a spectrum of materials commencing with **peat**, which is a surface deposit of decayed plant material. When buried to shallow depths, this becomes compressed into **lignite**, or 'brown coal', which is composed of carbon, plus 45–65% volatiles (mainly hydrogen, oxygen and sulphur). During further burial these gases are progressively driven off, resulting in what is termed '**bituminous coal**', which, with **anthracite**, are the main sources of fuel. Anthracite is the most efficient type of coal, being almost pure carbon, with a volatile content of 7–12%. Anthracite has been so deeply buried that it is effectively a metamorphic rock. Further metamorphism results in the mineral **graphite** (*see* chapter 7). Coal also contains small quantities (up to a few ppm) of impurities, some of which are potentially dangerous (e.g. mercury, arsenic and selenium).

Coal typically forms rock layers (strata) within deltaic sedimentary sequences of sandstones, shales and limestones. Carbonaceous deposits were formed throughout geological history, even in the Precambrian, but coal is especially associated with the Carboniferous Period, when vegetation flourished in low-lying swamps. The decaying plant material became trapped in mud, under acidic conditions, and protected from oxidation.

Uses

Coal is now mainly used as solid fuel for electric power generation; about 40% of the world's electricity is produced by this means. The coal is first pulverized, then heated, to produce steam to drive turbines, which in turn provide power for the generators. The residue from burning bituminous coal, known as **coke**, can be used as a reducing agent in iron smelting. Most modern plants, using anthracite, have an efficiency of around 45%. However, the combination of electricity generation and district heating in a new plant in Copenhagen gives a combined efficiency of 94% and offers the possibility of large emission reductions if widely copied.

In the process of coal gasification, coal is heated with oxygen under pressure to produce **syngas**, which is a mixture of hydrogen and carbon monoxide. Syngas can be employed directly as a fuel, or the hydrogen can be extracted for independent use; this will undoubtedly become an important power source for motor transport in the future. The combustion of syngas in a gas turbine is more efficient than burning coal, and the hot exhaust gases can be used in a steam turbine. This results in lower harmful emissions. Syngas is also employed to produce various hydrocarbon compounds, including methanol, acetic acid and ammonia, which in turn provide a range of chemical products. Nearly half of the syngas produced is destined for the manufacture of chemicals.

Coal can also be liquefied to give a synthetic fuel as an alternative to petroleum, and is becoming a significant alternative to solid fuel in China.

Peat is still exploited as a fuel in certain countries, and is also widely used as a soil conditioner.

Production

World production of coal in 2011 totalled 7695 million tonnes, the largest producers being China and the USA (table 9.2). There are very large reserves, both in some of the countries listed in table 9.2 and in several others, such as Zimbabwe, North Korea and Spain. World reserves were estimated at 860,938 million tonnes in 2012, although much of this is low-quality lignite.

country	million tonnes
China	3520
USA	992.8
Australia	5.8
India	5.6
Indonesia	5.1
EU*	4.2
Russia	4.0
South Africa	3.6

A. coal

country	thousand barrels/day
Russia	9934
Saudi Arabia	9760
USA	9141
Iran	4177
China	3996
Canada	3294
Mexico	3001
UAR*	2795
Brazil	2577
Kuwait*	2496
Venezuela	2471
Norway	2350
Iraq*	2400
Nigeria	2211
Algeria	2126

B. petroleum

Table 9.2 Coal production from main producing countries
Data: EIA International energy statistics, 2011.

Environmental concerns

The burning of coal is the biggest source of anthropogenic carbon dioxide release – 14,416 million tonnes of it in 2011 – double the emissions from gas-fired plants. Because of the dangers posed by global warming (*see* chapter 12), the UN Climate Agency advised in 2013 that most of the world's coal reserves should not be exploited. Nevertheless, the largest producers (China, the USA and India) are still using very large quantities, although both China and the USA are gradually reducing their consumption. In the case of the USA, this is due to the increasing use of shale gas as a cheaper alternative, whereas in China, coal liquefaction is gradually replacing the burning of solid fuel.

Both carbon dioxide and sulphur dioxide produce acid rain, which can acidify freshwater lakes and rivers with consequential effects on wildlife. However, the cooling effect of sulphur dioxide in the atmosphere to some extent counteracts the warming effect of the carbon dioxide.

Smoke from coal burning is a known cause of lung cancer, and the various waste products contained in coal ash residue are potentially dangerous, both as air pollutants, and as ground contaminants. These include mercury, arsenic, selenium, uranium and thorium. Underground coal fires are also a major environmental concern; these can ignite spontaneously or be started by lightning strikes or accidentally by human means. They can last for years; a fire in an anthracite mine in Pennsylvania started in 1961 and is still burning.

Oil and natural gas

The correct term for natural oil is petroleum, from the Latin *petra* (rock) and *oleum* (oil), which occurs in the form of either liquid oil, gas, or solid (i.e. asphalt – *see* chapter 8). It is a black or dark brown material consisting of a mixture of hydrocarbon compounds plus variable amounts of sulphur, nitrogen and hydrogen, and traces (less than 0.1%) of metals such as iron, nickel, copper and vanadium.

Origin

Petroleum is formed as a result of the partial decomposition of organic material – mostly in the form of planktonic marine organisms such as algae that have been deposited on the sea floor under reducing conditions, that is, cut off from oxygen. There the material is decomposed by bacteria to form an organic mud which, when buried and subjected to moderate heat and pressure, produces the petroleum.

The oil and gas is subsequently squeezed out of the source rock by the pressure of the overlying rock strata, and migrates upwards to accumulate in a suitable **reservoir rock** (Fig. 9.1). Oil reservoir rocks require to be both porous and permeable; that is, they must have open pore spaces to accommodate the oil and gas, and these spaces must connect together via channel ways. To contain the petroleum, the reservoirs must be sealed by a **cap rock** that is impermeable, such as shale or evaporite (e.g. salt or gypsum). Suitable geological structures that commonly form seals, known as 'traps', include anticlines, unconformities, salt

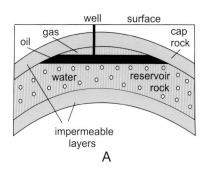

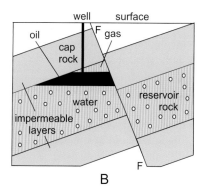

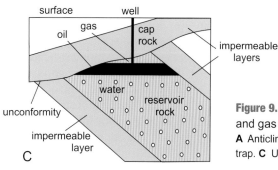

Figure 9.1 Structural traps for oil and gas
A Anticline trap. **B** Fault-bounded trap. **C** Unconformity trap.

domes and fault-bounded structures such as half-graben (Fig. 9.1). The oil 'floats' on saline water that fills the pore spaces underlying the petroleum within the reservoir; the gas accumulates above the oil. As the reservoir is depleted, the remaining oil becomes increasingly contaminated with water.

Most of the known petroleum reserves are held in reservoir rocks of Mesozoic or Cenozoic age, and at relatively shallow depths (less than 2.5km). Large oil fields are located in the Persian Gulf region, Central Asia, the southern USA, Alaska, Venezuela and West Africa. Significant deposits also occur around the British Isles, in the North Sea and on the Atlantic continental shelf. The latter deposits are derived from both Carboniferous and Jurassic source rocks and are now held in Mesozoic sandstone reservoirs.

History

Bitumen (asphalt) has been exploited since the earliest recorded times (*see* chapter 8); petroleum exploration, however, was first recorded in the fourth century CE by the Chinese, who used bamboo-drilled wells. In 1847, James Young in England invented the process of distilling **kerosene** from natural petroleum for use as lamp oil. However, the first commercial oil refining plant was established in 1851 at Bathgate in Scotland, utilizing oil shale and bituminous coal.

The first modern oil well in the USA was drilled in 1859, using a steam engine, and by the 1860s productive oil wells existed in several countries, including Romania, Canada and Azerbaijan. By the mid-twentieth century, the USA was the foremost producer but was soon overtaken by Saudi Arabia and the Soviet Union. These three countries are still the leading producers.

The first petroleum deposits were found by observing oil seeping out at the surface. Since that time, efforts to discover and develop oil fields have become ever more sophisticated. As the price of oil has increased, and as many of the older fields are nearing exhaustion, it has become economically more viable to search for and exploit much smaller or deeper fields.

Oil recovered directly from the source rock, known as **crude oil**, is very variable in quality, from 'heavy crude' to 'light crude' depending on its density, and also on the proportion of contained sulphur. Lighter crude oil is more desirable and thus more expensive.

Petroleum is increasingly being exploited, especially in the USA, from petroleum-bearing shales by hydraulic fracturing ('**fracking**'). In this technique, fluid containing solid particles such as sand is injected into the shales to induce cracks to open up in order to release the oil or gas.

Petroleum production

Crude oil is separated into various components by fractional distillation at an oil refinery, in which the lighter fractions are progressively driven off when heated (*see* table 9.3). The main products are **petrol** (called **gasoline** in North America), kerosene, natural gases such as **propane** and **butane**, **diesel** fuel, and other hydrocarbon compounds used in the plastics and pharmaceutical industries. Other products include lubricant oils, wax, aromatics, sulphur and asphalt.

World oil consumption is huge,

distillation fraction	boiling point ^{0}C
LPG	-40^0
butane	$-12 - 1^0$
petrol (gasoline)	$-1 - 110^0$
jet fuel	$150 - 205^0$
kerosene	$205 - 260^0$
fuel oil	$205 - 290^0$
diesel oil	$260 - 315^0$

Table 9.3 Distillation products of petroleum
Note: LPG = liquid petroleum gas.

about 87 million barrels[1] per day, the bulk of it for motor vehicle fuel, which consumes about 90% of the production. Oil is a key component of international commerce and its price is an important market indicator. Many of the more important oil exporting countries, mostly in the Middle East, have combined in a marketing organization (OPEC – Organization of Petroleum Exporting Countries) in an attempt to control the price; however, several major exporters, including the USA, Canada and Russia, are outside this organization, and the price is effectively determined by the open market. The top three producers in 2009 were Russia, Saudi Arabia and the USA (table 9.4).

Natural gas

Natural gas consists mostly of methane, with minor quantities of carbon dioxide, nitrogen and hydrogen sulphide. Its main uses are in domestic heating and cooking, electricity generation, and as a fuel for motor vehicles; it also

---

1 [1 barrel = 159 litres, or 42 US gallons]

country	thousand barrels/day
Russia	9934
Saudi Arabia	9760
USA	9141
Iran	4177
China	3996
Canada	3294
Mexico	3001
UAR*	2795
Brazil	2577
Kuwait*	2496
Venezuela	2471
Norway	2350
Iraq	2400
Nigeria	2211
Algeria	2126

Table 9.4 Petroleum production from main producing countries
Data: American Petroleum Institute, 2010.

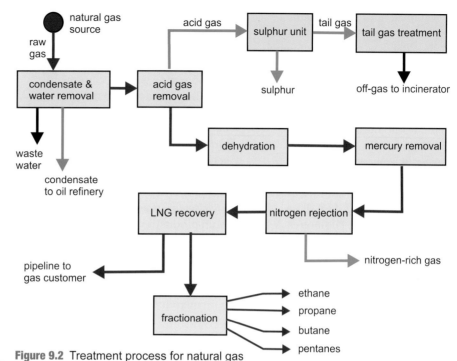

Figure 9.2 Treatment process for natural gas
Main products, red; by-products, green; waste products, black. Image credit: after Wikimedia Commons: NatGasProcessing.png.

provides a chemical feedstock in the production of plastics. The Chinese are known to have piped natural gas from natural seeps to provide a fuel for extracting salt from seawater, but the first modern industrial use of natural gas was in the USA in 1825.

Until the mid-twentieth century, gas obtained as a by-product from oil wells was flared off at source but is now increasingly being piped separately for industrial or domestic use, or in some cases returned to the reservoir to aid the oil recovery.

Natural gas can be exported directly as liquefied natural gas (LNG) for power generation; it can be converted into synthetic petrol (gasoline) or processed to produce various other petroleum products such as ethane, propane or butane; impurities such as hydrogen sulphide, carbon dioxide, nitrogen and water may need to be removed (Fig. 9.2).

Natural gas is particularly useful as a fuel in combination with renewable-energy power sources in electricity generation as a back-up, because of its ability to cope quickly with changes in output. It is preferable to oil or coal because it produces significantly less carbon dioxide emission per unit of energy (30% less than oil and 45% less than coal) and is increasingly being advocated for modern power stations. It can be used to produce hydrogen as a fuel for motor vehicles, and ammonia for the production of fertilizers. The gas is transported via a network of pipelines on land, but is shipped in the form of LNG by sea.

Biogas and gas hydrates
Methane gas is produced naturally (i.e. '**biogas**') as the result of anaerobic decay in swamps; it is also generated artificially from landfill sites, agricultural waste and sewage sludge and, after removing contaminants, can be exploited in biogas reactors to generate power. This source has the potential to be greatly expanded, thus saving on conventional fuels, and has the additional advantage of eliminating a potent greenhouse gas from the atmosphere.

Solidified (crystallized) natural gas, known as **gas hydrate**, occurs in offshore deposits on the continental shelves and in the arctic permafrost. The warming of the latter is a serious environmental concern because of the potentially damaging release of methane into the atmosphere.

Petroleum reserves

There are large known reserves of oil: about 190 cubic kilometres (1.2 trillion barrels), and if oil shale is included, 595 cubic kilometres (3.74 trillion barrels) – enough to last for 120 years at present rates of usage. However, only a small proportion of these are recoverable by present methods, and the international concern over the projected environmental damage that would be caused by extracting and burning this quantity of petroleum means that most of these reserves will probably never be exploited. Oil is increasingly being replaced as a fuel by coal (which is even more damaging) or the more environmentally friendly alternatives of nuclear power and renewable energy sources such as solar, wind, tidal and wave power (*see* chapter 11).

Natural gas reserves are also considerable; in 2013 it was estimated that there were 33.6 trillion cubic metres in Iran, 32.9 in Russia, 25.1 in Qatar, and 17.5 in Turkmenistan. Shale gas reserves amounted to a further 180 trillion cubic metres.

Environmental concerns

Petroleum is a major cause of environmental concern for several reasons. Oil spills, especially at sea, are a serious hazard to the marine environment; large spills such as the sinking of the Exxon Valdez oil tanker off Alaska, and the more recent Gulf of Mexico episode, caused widespread destruction of marine wildlife. Natural oil spills have always existed and are not a serious problem, since the oil is naturally degraded in time by bacteria, but the increased scale of the larger accidental spills does create a significant hazard. Contamination both from methane and from chemicals added to fracking fluids are a concern in petroleum production employing the fracking process, and have been the cause of much public opposition to the introduction of this technique in the UK.

Gas emissions are dangerous – both methane and hydrogen sulphide are toxic; methane is also odourless, although artificial odours may be added for safety. Explosions caused by leaks, both of domestic gas and at industrial plants, are not uncommon.

A much greater concern, however, is the environmental damage caused by the burning of fossil fuels, which releases large quantities of the so-called 'greenhouse' gases, chiefly carbon dioxide, into the atmosphere. This problem is discussed in more detail in chapter 12.

Nuclear power

Nuclear power could be regarded as a somewhat borderline non-renewable resource, since it is often grouped with the renewable energy sources because of its 'green' credentials as a zero-carbon emitter. Although, in theory, nuclear power systems could operate with minimal demands for fresh fuel, working nuclear power plants use significant amounts of uranium, which is, of course, a non-renewable resource, albeit that supplies are sufficient to last for many centuries.

History

Nuclear power as a source of energy arose from the development of the nuclear bomb in the 1940s. The technology is based on the fact that heavy atoms such as uranium can be split into lighter atoms (i.e. by **nuclear fission**) by bombarding them with neutrons, thereby releasing large amounts of energy, plus additional neutrons which, in turn, produce further fission. This process forms a **chain reaction** which, if uncontrolled, causes the kind of explosion seen in the atomic bomb, but if controlled is an important source of energy.

The first nuclear reactor was built in the USA in 1942 as part of the 'Manhattan Project' and produced enriched uranium and plutonium for the first atomic bombs. However, the first nuclear power plant to produce electricity for the national grid was made in the USSR in 1954, and was shortly followed by others in the UK and the USA. Thirty-one countries now operate nuclear power plants and around 6% of the world's energy comes from nuclear sources. The seven leading countries using nuclear power are listed in table 9.5.

Nuclear reactors are also used as propulsion units for military purposes, especially in submarines and large surface vessels.

Technology

The basis of nuclear power is relatively simple: the nuclear reactions just described, being exothermic, release

country	capacity kilo-watts	% total electricity	nuclear reactors
USA	99,081	19.4	100
France	63,130	73.3	58
Japan	42,388	17.1	48
Russia	23,643	17.5	33
South Korea	20,721	27.6	23
China	17,978	2.1	21
Canada	13,538	16	19

Table 9.5 Nuclear power in top seven producing countries
Capacity in kilowatts, percentage of total electricity generated, and number of nuclear reactors. Data: International Atomic Energy Association, 2010.

heat, which is removed by a cooling system and used to generate steam to power a turbine; this drives the generator supplying the electricity. The process begins with uranium ore (chiefly pitchblende – *see* chapter 7). The uranium oxide is converted to fluoride, enriched and made into the fuel rods, which are the basis of the reactor. These last for about six years, after which the 'spent' fuel rods are removed and can be reprocessed; in theory, 95% of the fuel can be re-used. After about five years, the spent rods have lost much of their short-term radioactivity and are cool enough to move to a storage facility or to be reprocessed.

Fuel resources
Reserves of uranium have been calculated to last at present rates of usage for at least several centuries. However, if reprocessing techniques are improved and expanded, the reserves are probably enough to last for many thousands of years. Moreover, improved reactor designs using the common isotope U238 – the so-called 'fast-breeder' reactors – are being tried in France and Russia, but are not economic at present because the comparative cheapness of uranium favours the present generation of reactors, compared with the expense of investing in the newer technology.

Disadvantages
There are three important disadvantages to nuclear power: the high cost of building and decommissioning the power plants; the problems and cost of the disposal of the radioactive waste material (including those associated with decommissioning) and the serious consequences of accidents. The building costs compare unfavourably with modern gas-fired power plants, but there are no ongoing fuel costs and the nuclear power stations can remain operable for longer.

Public opinion understandably has focused on the dangers posed by accidental releases of radioactivity. There have been several catastrophic accidents that have released dangerous quantities of radioactive material into the atmosphere, and around 100 reported 'minor' accidents. The Chernobyl accident in 1986, which completely destroyed the power plant, is thought to have caused around 60 immediate deaths plus many thousands of latent cancer deaths, but it also poisoned grazing land in much of western Europe: the hill farms in Wales and Cumbria (NW England), which are subject to particularly heavy rainfall, were only cleared to market their sheep in 2012. The Fukoshima disaster in 2011 displaced 50,000 households and the fallout of radioactive particles contaminated the surrounding area, from which vegetables and fish were banned. Although there are fewer accidents than at conventional power plants, those at nuclear plants are inherently more serious; the possibility always exists that human error or a breakdown of the cooling system will cause a controlled nuclear reaction to become uncontrolled, resulting in the meltdown of the reactor core, as happened at Chernobyl.

Waste disposal is also a serious problem. Spent fuel consists of unconverted uranium and plutonium plus the radioactive fission products of the reactors – the so-called 'high-level' waste. To this is added large quantities of 'low-level waste' consisting of contaminated clothing and tools, together with the building materials from decommissioned reactors. Some of this waste material remains radioactive for hundreds or thousands of years and must be adequately shielded from the biosphere for the indefinite future. Most are currently held in repositories at the site of the power plants, but several countries aim to build deep underground ('geological') repositories for the high-level waste in suitable rock judged to be unsusceptible to leakage or earthquake damage. However, the transport of the waste to a central location introduces its own problems, as there is considerable public opposition; overland transport of nuclear waste is banned in the USA.

Reprocessing of the spent nuclear fuel can theoretically recover 95% of the uranium and plutonium and convert it into usable fuel, with a consequent reduction in the long-term radioactivity of the waste. This is done at government level in the UK,

France and Japan, but is expensive and not commercially viable.

The opposition to nuclear power as a whole has been influenced by a strong international anti-nuclear movement, mainly against its use in weapons. The main nuclear powers are committed by treaty to a policy of opposition to the spread of nuclear weapons capability to other countries and to the reduction of their own stockpiles of nuclear weapons.

Future use

While there is no doubt that improved techniques can greatly reduce the dangers posed by accidents and waste disposal, the nuclear industry will always be potentially hazardous, and public concern in several countries has led to the reduction or abandonment of their nuclear power programmes. The use of nuclear power, which rose rapidly from the 1960s, reached its maximum midway through the 1990s and a number of planned plants have since been cancelled, especially since Fukushima (Fig. 9.3). Germany has been forced by public opinion to shut down all its reactors by 2022.

Nuclear fusion technology, which is potentially much safer, though technically more difficult than fission, is being investigated in several experimental reactors which, however, are not expected to begin operation for at least a decade. Thorium is currently being investigated as a safer fuel source than uranium because its radioactive breakdown products have relatively short half-lives and there is virtually no danger of a runaway reaction causing meltdown.

However, despite its obvious attractions as a reliable low-carbon power source, it seems likely that nuclear power will be replaced in popularity, to a large extent, by the various renewable energy sources described in chapter 11.

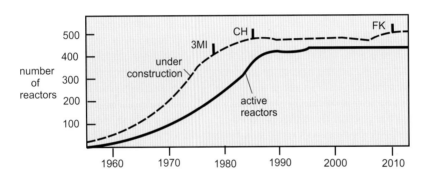

Figure 9.3 World nuclear power production, 1960–2010.
Growth history of the nuclear power industry over the last 50 years. Note the effects on nuclear plant construction of the three major disasters, at Three-mile Island (3MI), Chernobyl (CH), and Fukushima (FK).

10   Atmosphere, ocean and biosphere

The atmosphere and the oceans form the most valuable resource that the Earth possesses; they provide the environment within which all forms of life exist, and without them, life as we know it could not survive. The **biosphere** is the name given for this global system together with all the living organisms that exist within it.

The atmosphere

The present composition of the atmosphere bears little relationship to its original composition, being very largely a product of its interaction with biological processes over geological time (*see* chapter 2). It is thought that the primordial atmosphere would have been lost during the initial stages of Earth's existence and would have been replaced by one derived from volcanic exhalations and comet impacts, rich in carbon dioxide and water, and with significant amounts of the reducing gases hydrogen, methane and ammonia. For the first two billion years or so, this atmosphere sustained forms of life such as bacteria that relied on photosynthesis and consumed carbon dioxide to provide energy; oxygen was excreted as an unwanted waste product. The atmosphere gives some protection to living organisms from genetic damage by the ultra-violet solar radiation. It also prevents excessive destruction from meteorites, most of which burn up as a result of friction as they pass through it. The contrast with the surface of the moon, which is unprotected by an atmosphere, is revealing (*see* Fig. 2.6).

The atmosphere is a vital resource, which is often taken for granted. Yet changing its delicate balance of gases could have catastrophic effects. It is important, therefore, to understand the mechanisms that have controlled this balance over geological time so that appropriate policies can be put in place to limit or reverse the dangerous effects of human interference on atmospheric composition.

Structure

The lowest and most dense layer of the atmosphere is termed the **troposphere** (*see* Fig. 2.5B) which extends upwards to about 12km, on average, above the surface. It contains the mixture of gases that we know as the 'air'. Both the density and the temperature of the air in the troposphere decrease upwards – in the case of temperature, at a rate of about 1°C every 100 metres. Because warmer air is less dense than cooler air, it rises, then cools at a higher level. The circulation pattern that causes what we call 'weather' occurs because of thermal differences between different areas of the Earth's surface. Thus hotter air at the Equator is transported to higher latitudes as a result of thermal convection, and because cooler air can hold less water, the excess water condenses into clouds and ultimately results in rain.

Although very complex in detail, the movement of air horizontally across the surface of the Earth, which we experience as wind, is ultimately controlled by warmer and less dense air moving upwards being replaced by cooler and denser air flowing near the surface. The surface wind directions and velocities are controlled by pressure differences, flowing from the denser high-pressure areas to less dense low-pressure ones. These movements occur in a series of cells whose mean positions are shown in Figure 10.1.

The prevailing surface wind directions are affected by the rotation of the Earth, which imparts a component of latitudinal movement; hence, for example, the south-westerly prevailing winds in the northern temperate zone. Disturbances to this simple pattern are produced by the interactions of the colder and warmer air, which produce the rotating, roughly circular, concentrations of rising hot air known as **cyclones** (Fig. 10.2). In extreme cases, these produce the severe weather events known as hurricanes and typhoons; the winds around these circulatory systems can reach over 250km per hour and cause massive damage. These cyclonic systems are separated by the areas of high pressure termed **anticyclones**. Because these weather systems are sensitive to quite small changes in average sea temperature, global warming is predicted to cause a significant increase in both their frequency and their severity.

Above the troposphere is the **stratosphere**, which is about 40km thick

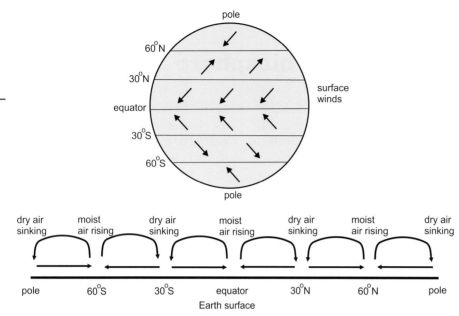

Figure 10.1 Prevailing wind directions
Warm moist air rises at the Equator and sinks at the tropics, replaced by drier surface air flowing towards the Equator. A similar cycle occurs in the temperate zones: moist air rises around 60° latitude and sinks at both the tropics and the poles, replaced by the return flow of colder, drier air. The effect of the Earth's rotation (Coriolis effect) causes the prevailing winds to flow in a NE or SW direction in the northern hemisphere and NW or SE in the southern.

and hosts very strong winds known as the **jetstream**. Because of the low density of air molecules, the effects of solar radiation are much stronger, and consequently the temperature in the stratosphere increases upwards, in contrast to that in the troposphere. Hence the air is very stable, and little mixing takes place. The lower part of the stratosphere contains the **ozone layer**, where solar radiation converts the oxygen molecules into ozone (O_3).

Above about 50km from the surface are two further layers, the **mesosphere** and the **thermosphere** (*see* Fig. 2.5B). The air here is very thin, and the unfiltered sunlight accelerates the rate of chemical reactions; gas molecules are broken down to form positive ions and negative electrons, resulting in the electrically conducting layer known as the **ionosphere,** which is responsible for the **aurora borealis** and **aurora australis**. The thermosphere grades up into the **exosphere**, which gradually merges upwards into the rarefied atmosphere of the outer Solar System.

Composition
The present composition of the atmosphere consists of nitrogen (78%), oxygen (20%), argon (0.9%) and carbon dioxide (0.03%) together with trace proportions of other gases: methane (100ppm), nitrous oxide (10ppm), ammonia (1ppm), methyl chloride (0.1ppm), sulphur compounds (0.001ppm) and methyl iodide (0.00001ppm). The proportion of water is highly variable but averages around 1%.

The nitrogen molecule is strongly bonded and difficult to break down, so the nitrogen in the atmosphere is very stable. Nitrogen is released by bacteria and absorbed by sea water in the form of nitrates.

Oxygen is produced by photosynthesis in plants and is consumed by the respiration of animals and by other processes such as the burning of vegetation. Oxygen is also released when carbon compounds from decaying organic matter are buried, or washed into the sea. The present proportion of oxygen is close to the upper safety limit for much planetary life. It has been calculated that even a small increase in the oxygen level would render vegetation much more inflammable, with the result that lightning-induced fires would ultimately destroy much of the vegetation cover.

Since carbon dioxide is emitted by all air-breathing animals and consumed by plants, its proportions in the atmosphere are strongly influenced by the biosphere. It is also partly controlled by its reaction with sea water. There is about 50 times as much carbon dioxide dissolved in the ocean in the form of bicarbonates as there is in the atmosphere. Small quantities of carbon dioxide are dissolved by atmospheric water to produce weak carbonic acid.

Methane is also a product of the biosphere; about 500 megatonnes of methane are produced annually from bacterial breakdown in anaerobic conditions in the muds of the sea bed and in marshes. In the stratosphere, methane

Figure 10.2 Tropical cyclone
Cyclone 'Nathan' approaching the coast of northern Queensland, Australia, 19 March, 2015. Note clockwise sense of rotation. Image credit: NASA image.

oxidizes into carbon dioxide and water; the water dissociates into hydrogen, which escapes into space, and oxygen, which descends into the troposphere. The oxidation of methane uses up large quantities of oxygen that might otherwise rise to dangerous levels.

Nitrous oxide, although produced in considerable quantities, constitutes an even smaller proportion of the atmosphere than methane; this is because it is destroyed by ultraviolet sunlight, releasing nitrogen and oxygen. Thus, to some extent, nitrous oxide and methane are complementary, in that the one releases oxygen whereas the other consumes it.

Around 1000 megatonnes of ammonia are produced annually from biological activity and is important in counteracting the acidity of rainwater.

Sulphur dioxide is produced naturally from vulcanicity and also from the burning of fossil fuels. Rainwater containing weak sulphuric acid can be toxic to wildlife in freshwater lakes and rivers, as has been shown to be the case in parts of Scandinavia and North America.

The oceans

Like the atmosphere, the oceans represent a vital Earth resource. The world ocean occupies over two-thirds (71%) of the Earth's surface and consists of five separate but linked bodies: the Pacific, Atlantic, Indian, Southern and Arctic Oceans. It contains 97% of the Earth's water with a total volume of 1.3 billion cubic kilometres. The ocean controls weather and climate and is an essential part of the biosphere. The source of the water is rather mysterious; it is not thought to be an original component of the primeval Earth, but it has been suggested that it may have arrived early in the Earth's history from extra-terrestrial bodies such as asteroids and comets.

Composition
Ocean water contains small quantities of a number of elements and molecules in the form of ions in solution. The most important are carbon dioxide (64–107 ppm), nitrogen (10–18ppm) and oxygen (0–13ppm). There are also significant amounts of iron, aluminium, manganese, silicon, calcium, sulphur, potassium, magnesium, sodium and chlorine. The salinity is quite variable and depends on temperature; the waters in polar regions have a lower salinity, since precipitation there exceeds evaporation. The opposite is the case in the tropical and temperate regions, where evaporation dominates.

Structure
The average depth of the ocean is 3790 metres but the deepest parts, the deep ocean trenches, are between 6000 and 11,000 metres in depth. The uppermost layer of the ocean, where photosynthesis takes place, is termed the **photic zone** and extends to depths of up to 200 metres. The photic zone forms part of a thin surface layer, containing only about 2% of the ocean water, which is well mixed due to the continual action of waves and currents and is characterized by uniform salinity and temperature. It is this layer that is in contact with the atmosphere, and which initiates the world's weather. Beneath this surface zone is a narrow transitional layer known as the **thermocline**, lying between 700 and 1000 metres depth, marked by a sharp change in both temperature and salinity. Here the density increases sharply with depth as the temperature decreases. The thermocline lies at a deeper level in tropical regions

than in more temperate waters, and is absent altogether in the polar regions, where the water is uniformly cold.

Below the thermocline lies the deep zone, which contains the bulk of the ocean water (c.80%), and which is colder, denser and more stable than the upper layers. The density here shows little change with depth.

Ocean currents
The ocean is constantly moving water around the Earth's surface by means of a permanent system of currents (Fig. 10.3). These affect the climate by transferring heat from the tropics to temperate and polar regions, where evaporation of the warmer and saltier water results in the precipitation of rain or snow. This process is balanced by the transfer of cooler, denser and less salty polar water towards the tropics. These currents only affect the surface zone, and are partly driven by the prevailing winds. They have been known and used by

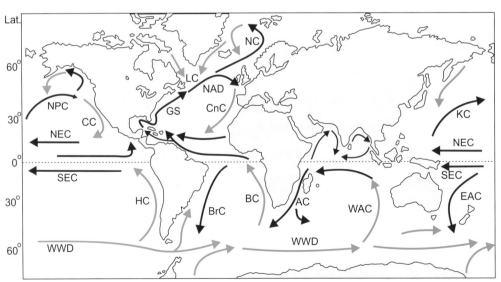

Figure 10.3 Oceanic surface current pattern
AC, Aguilhas current; BC, Benguela current; BrC, Brazil current; CnC, Canaries current; CC, California current; EAC, East Australia current; GS, Gulf Stream; HC, Humboldt Current; KC, Kuroshio current; LC, Labrador current; NAD, North Atlantic drift; NC, Norwegian current; NEC, North Equatorial current; SEC, South Equatorial current; WAC, West Australia current; WWD, West Wind drift.

mariners for centuries, especially in the days of sail, in order to choose the easiest routes across the oceans.

The currents in the Northern Hemisphere below about latitude 45°, form a clockwise circulation pattern, whereas those of the Southern Hemisphere move in an anticlockwise sense (Fig. 10.3). This has the effect of making the climate of the north-eastern shores of the North Atlantic warmer than equivalent latitudes of the western, for example. Thus the **Gulf Stream** makes the climate of Northwest Europe much warmer than that of eastern Canada, which is affected by the cold **Labrador Current**. Similarly, in the Southern Hemisphere, the cool **Humboldt Current** makes the western coast of Peru much cooler than the east coast of Brazil, which is warmed by the **Brazil Current**.

There is also a set of deep ocean currents, which are caused by the down-welling of colder and denser polar water and which circulate at depth, largely unconnected to the surface currents. They move along the western sides of the ocean basins at a much slower rate than the surface currents. These deep, cool currents are rich in nutrients and plankton, and where they rise up in the equatorial regions, they support a rich fauna.

Global warming poses important environmental problems for the ocean current system, having the potential to warm the Arctic to the point where the supply of cool polar water is reduced or disrupted. This could have the effect of diverting the Gulf Stream away from northwest Europe, possibly resulting in a climate more typical of eastern North America.

The biosphere

The **biosphere** is the name given to the global ecological system containing all forms of life, together with the environment in which they live, whether that be the oceans, the land or the atmosphere. It has evolved continuously from the earliest appearance of life nearly four billion years ago, gradually adapting to the various geological changes that have affected the planet. The totality of all living organisms is known as the **biomass**.

Life forms have adapted to the most extreme conditions that the planet has presented. Micro-organisms have been detected at heights of 40km in the stratosphere, at depths of over 10km in the deep ocean trenches, and in boreholes within the crust at depths of around 5km. Some have adapted to extreme cold in the polar regions, and intense heat in hot volcanic springs on land and around black smokers in the deep ocean. Although there have been several mass extinctions of species at times of abrupt geological change (*see* below), enough have always survived to provide a new explosion of life forms when conditions improved.

The composition of the present biosphere has been drastically changed as a result of human intervention. The atmosphere has been altered by the addition of greenhouse gases and the ocean has been polluted by various substances that are dangerous or even toxic to marine life. However, the biggest change to the biosphere has been brought about because of food production by an increasingly populous human race. Forests have been cleared, soil has been eroded, and vast land areas have been modified by agricultural activity, with the result that the original plant and animal inhabitants have been replaced by new ones. Thus the composition of the present biomass is very different from that which existed before humans began to cultivate the land.

Biodiversity

The extent of variability in the biomass is known as its **biodiversity**, and can be measured by the number of different species within a given area, or on the Earth as a whole. It varies according to the local environment, being greatest in equatorial regions and least at high latitudes. Rainforests exhibit the greatest diversity; Brazil's Atlantic forests have around 20,000 plant and 1350 vertebrate species, together with several million insect species. Land masses that separated during the break-up of **Pangaea** during the Mesozoic Era (e.g. Australia, South America and Africa) have evolved quite different suites of organisms, which have contributed to the Earth's overall biodiversity. Unfortunately, however, the introduction of alien species into these separate ecosystems has had the effect of reducing their natural biodiversity, by out-competing with the native flora or fauna. The effect of introducing the rabbit to Australia is a well-known example of this.

The Earth's biomass is essential to our own survival as a species, since it provides all our food requirements. Natural biodiversity is a valuable resource, since it provides a range of possible genetic variants from which to select new, improved types of plant or animal. However, the natural biodiversity from which humans have selectively bred useful variants over the millennia has been drastically

reduced; it is currently threatened by habitat destruction (e.g. clearing of rainforest), pollution, over-intensive cultivation and the effects of invasive alien species. A recent estimate has predicted that, at the present rate of extinction, 30% of all species will have disappeared by 2050; this is hundreds of times faster than the natural (geological) species extinction rate.

Climatic variation through geological time

The Archaean

The geological record during most of the Archaean shows no evidence of glaciations, despite the fact that the Sun then emitted only about 75% of its present heat. The additional warmth then is most probably to be explained by the greenhouse effect of much higher levels of carbon dioxide than today's, perhaps by a factor of 20, together with methane, which, in the absence of oxygen, would not have been oxidized to carbon dioxide as occurs today. Ammonia may also have been present; although it is now broken down to nitrogen and hydrogen by ultraviolet light, this process may have been inhibited then by a much denser atmosphere.

The earliest recorded glaciation occurred towards the end of the Archaean, about 2.9 billion years (2.9Ga) ago, from the evidence of glacial boulder clays. These were probably confined to the polar regions, but later glaciations, at 2.45Ga and 2.3Ga, seem to have been more widespread, the latter having reached tropical latitudes. Red beds appear for the first time in the geological record around 2.4Ga ago, indicating that there was enough free oxygen in the atmosphere to oxidize the iron in sediments. This event is thought to have created the ozone layer. It is possible that it was the presence of oxygen in the atmosphere that, by oxidizing methane, removed so much of it that the consequent cooling initiated the 2.45 glaciation.

The Proterozoic

During the Early Proterozoic, the Earth had warmed up, and by about 1.8Ga ago the first eukaryotic cells (containing nuclei) appeared, leading to an increase in the complexity of life forms; this may only have been possible owing to the presence of sufficient oxygen.

Three separate glaciations are recorded in the Late Proterozoic: at c.740Ma, 660–635Ma and 580Ma, each probably lasting for several million years. It has been suggested that the 740Ma event may have been triggered by the break-up of the supercontinent of **Rodinia** at around 800Ma into a number of smaller landmasses. This would have greatly increased the global extent of orogenic mountain belts and the length of shorelines. The effect that this would have had on the extent of silicate weathering and the consequent removal of carbon from the atmosphere is discussed below.

The ending of the Late Proterozoic glaciation coincided with the emergence of multicellular life, which appeared at around 580Ma ago and culminated in the explosion of complex organisms that mark the beginning of the Cambrian Period 40Ma later. The removal of the ice and the increase in land-derived nutrients could have encouraged the growth of algal micro-plankton that would have absorbed much of the dissolved carbon dioxide, leaving the atmosphere enriched in oxygen.

The Phanerozoic

The pattern of global temperatures since the late Archaean, consisting of long stable periods of relative warmth punctuated by relatively short glacial events, continued into the Phanerozoic, which contained three major glacial intervals; the first, spanning the Ordovician–Silurian boundary, at c.440Ma BP, the second, during the late Carboniferous and early Permian, at 325–280Ma BP, and the third commencing in the Paleogene at c.40Ma ago and continuing until the present (Fig. 10.4). Each of these glacial periods consists of a number of glacial peaks separated by warmer intervals, such as the present one, and each produced at least one large polar continental ice sheet. In contrast to the Precambrian glaciations, the Phanerozoic **tillite** beds are not overlain by carbonates, suggesting that temperatures did not fluctuate as violently as they had during the Precambrian.

Snowball Earth

It has been suggested that once complete ice cover reaches within 30° of the Equator, the ice cover would inevitably spread over the whole Earth. This phenomenon, christened '**Snowball Earth**', is thought to occur because ice and snow reflect much of the sunlight back into space (i.e. they have a high **albedo**), whereas land surfaces and ocean water absorb the Sun's heat. Thus, the greater the extent of ice and snow cover, the more the Earth's surface cools. This is a positive feedback mechanism and should tend to lead inevitably to complete snow/ice cover of both land

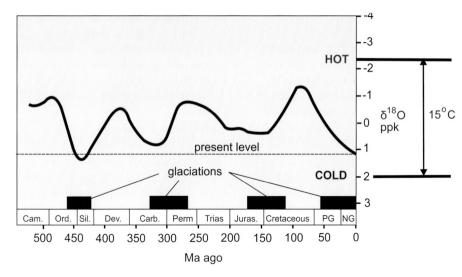

Figure 10.4 Temperature variation over the last 500 million years
Simplified curve showing temperature changes from the Cambrian Period (540Ma ago) to the Present, derived from oxygen isotope ratios ($\delta^{18}O$ parts per thousand).

and ocean – the snowball Earth – if not countered by some other mechanism. In a snowball Earth, oxygen in the oceans would gradually be used up, no sunlight could have penetrated the thick ice cover, there would have been no evaporation from the oceans, and consequently no rainfall, erosion or sedimentation. However, if this had indeed occurred during the geological past, the fossil record would have shown a mass extinction of almost all forms of life, whereas, in fact, many types of organism do survive these glacial periods. Moreover, many of the preserved glacial deposits grade into normal water-laid sediments, indicating the continuing presence of sea water. Thus the tendency to progress towards a complete Snowball Earth must have been countered by some other process or processes that provided negative feedback. Some of these are discussed below.

Controls on global temperature
The greenhouse gases

Heat is transported around the Earth by the movement of vast masses of air and water through the process of thermal convection. Although both water and air are transparent to the visible spectrum of sunlight, infrared light is absorbed by a small number of gases that are present in trace quantities. These are the **greenhouse gases**, so-called because they trap part of the Sun's heat within the atmosphere, preventing it being radiated back into space in the same way that a greenhouse is warmed above the ambient temperature purely by the action of sunlight through the glass roof.

The most important greenhouse gases are carbon dioxide and water vapour; methane, although present in much smaller proportions, is also significant since it is about 20 times more potent than carbon dioxide. They also include tiny quantities of gases released through human activity, such as the notorious CFCs (chloro-fluoro-carbons), held responsible for the partial destruction of the ozone layer above the polar regions.

The significance of the carbon cycle

Rainwater dissolves carbon dioxide from the atmosphere, thereby becoming weakly acidic. Weathering and erosion of silicate rocks by this weak acid break down the silicates of metals such as calcium and magnesium to form soluble bicarbonates, which are carried into the oceans, resulting in the transfer of carbon from the atmosphere to the oceans. The calcium and carbonate ions are then absorbed by marine organisms to make their shells, and when these die, they form carbonate deposits on the sea floor. These, in turn, are affected by plate-tectonic processes; subduction-related vulcanicity liberates carbon dioxide back into the atmosphere, thus completing the cycle. Elsewhere, orogenic uplift elevates the shallow-marine carbonates above sea level, where they are weathered and eroded again, returning the carbon to the ocean.

However, when carbon dioxide levels in the atmosphere increase, causing the global temperature to rise, there is greater evaporation of sea water, increased rainfall, and consequently more silicate weathering and more carbon transfer to the oceans. Silicate weathering thus acts as a negative feedback mechanism countering global warming. The opposite holds true when global temperatures are low; there is less rainfall, slower silicate weathering, less carbon returned to the oceans, and hence more carbon dioxide retained by the atmosphere.

Glacial deposits in the geological record are often found along with carbonates, which always post-date the glacial sediments. These must represent the recovery of the planet from glacial conditions, since carbonates are usually associated with tropical waters, and suggest that the circulation of carbon from carbon dioxide to carbonates via the silicate weathering mechanism is the key to recovery from a global cooling event. It would appear that as the ice-snow cover extended over the Earth's surface, evaporation of ocean water would have been restricted, allowing carbon dioxide in the atmosphere to increase. When carbon dioxide levels reached a critical level, the cooling event would go into reverse, the snow-ice cover would begin to melt, and the excess carbon would be deposited as carbonates. However, the feedback mechanism works too slowly to counteract the excessive release of carbon caused by human activity since the Industrial Revolution, hence the danger of a more rapid and uncontrolled global warming.

The role of plate tectonics
Another reason why the Earth is not in a permanent snowball state must also lie, at least partly, in the plate-tectonic mechanism, which is internally driven, unaffected by global cooling, and continually supplies vulcanicity, which must locally penetrate the ice cover. This would keep parts of the ocean warm enough to sustain life and supply evaporation. Moreover, the clouds of volcanic ash that result from major eruptions would provide

a thermal blanket to counteract the cooling effect of the glaciation.

The reason why glaciations commence at all in an Earth where a warm greenhouse is the default state (Fig. 10.4) may also be due to plate-tectonic activity. It is probably not coincidental that the major cooling events of the Late Proterozoic and the Phanerozoic were all preceded by periods of continental amalgamation resulting in a global increase in the extent of collisional mountain belts and, consequently, of silicate weathering, which would be likely to remove greater amounts of carbon dioxide from the atmosphere.

The role of ocean currents
An additional factor that must influence global warming or cooling is the way that the oceans transfer heat internally by means of the ocean currents that circulate warm or cool water globally (see Fig. 10.3). One result of a partial cooling event that affected a large proportion of the Earth's surface would be to reduce the extent of this oceanic circulation and tend to concentrate the warmer waters near the Equator, thereby counteracting the effects of the global cooling.

The role of methane
Although methane is present in the atmosphere in much smaller proportions than carbon dioxide, it is a much more potent greenhouse gas. Because it is produced by the breakdown of organic material by bacteria, there must be a connection between the worldwide extent of such activity and the amount of methane in the atmosphere. Global warming events may cause excessive release of methane from marshlands

and permafrost, thus enhancing the warming effect; conversely, during glacial periods, falling sea levels may expose seabed methane hydrate deposits to erosion, which also releases the methane, counteracting the cooling.

Goldilocks and Gaia
In the well-known children's story *Goldilocks and the three bears*, Goldilocks successively tries the chairs, porridge, and finally the beds of Father, Mother and Baby Bear, only to find that only those of Baby Bear were just right – not too hard or too soft, or too hot or too cold, and so on. Earth has been referred to as the 'Goldilocks planet' because, in contrast to the other planets of the Solar System, conditions are just right for life to exist. None of the other solar planets are suitable; of our nearest neighbours, Mars is too cold[1*] and Venus too hot. Earth is the only solar planet to experience continuous plate tectonics and vulcanicity, and, crucially, to have contained surface water up to the present. Not only are conditions just right now, but they have been for over three billion years, despite wide variations in temperature and large changes in atmospheric composition. The surface was never too warm, and even during the most severe glacial periods, never too cold; negative feedback mechanisms have always been sufficient to bring the planet back to its default state of surface warmth, averaging between around 10 and 22°C. Because of the complexity of the ocean–atmosphere system, and

1 [However, the recent discovery of methane on Mars suggests that organisms may well exist, or have existed, there.]

of the many factors that influence it, there is no generally agreed mechanism held responsible for keeping the Goldilocks planet in balance.

Gaia

The concept of a self-regulating Earth, which he named **Gaia**, was originally proposed by James Lovelock in 1979. His proposition was that the Earth was kept in a habitable state essentially by biological means. That is, the organisms of the biosphere have always acted in such a way as to ensure that the Earth was suitable for their survival.

Although the present composition of the atmosphere has certainly been determined by the actions of organisms over geological time, it is a much more challenging concept that the biosphere actually controls the atmosphere in order to maximize its own survival chances – and impossible to prove conclusively. It is probably the case that, for example, excessive amounts of carbonate in the oceans could be countered by an expansion of carbonate-using shelly organisms, and that excessive quantities of oxygen in the atmosphere could be used up by additional animal activity or by an increase in forest fires. However, it would be difficult, and certainly beyond the scope of this book, to prove that biological regulation, rather than the more physical controls discussed earlier, was the ultimate determining factor in keeping the planet habitable.

11 Renewable energy resources

The main types of renewable energy resources are: biomass (organic material), solar radiation, geothermal, wind, hydraulic, wave, and tidal power.

Biomass

Biomass in the form of wood is the oldest form of energy source, having been used since the earliest times as a fuel for heating and cooking. However, the term 'biomass' covers any kind of biological material derived from living (or recently living) organisms; long-dead organisms (e.g. in the form of coal or oil) are, of course, a non-renewable fossil fuel resource. Biomass can be burnt directly to provide heat, or converted into fuel by some indirect chemical process.

The production of **charcoal**, obtained by the slow burning of wood in the absence of oxygen, was widely practised, especially for iron smelting and metal working, until replaced by coal during the Industrial Revolution. However, this was extremely wasteful and resulted in widespread deforestation. Modern methods of production of briquettes or pellets using various types of wood by-products are much more efficient.

Thermal conversion

Direct burning of biomass is very inefficient, converting only 7–27% of its potential energy. However, combined with coal, much higher efficiencies of 30–40% are achieved. The major objection to this use of biomass is its environmental effects – the release of not only carbon dioxide, but also other pollutants such as nitrous oxide and particulates. The dense industrial fogs ('**smogs**') that formerly blighted large European cities have thankfully now disappeared owing to the diminished use of coal for domestic fuel, but are still a serious problem in many countries such as China.

Biomass can also be burned in a furnace to provide a combustible gas, termed '**producer gas**', which is composed mainly of carbon monoxide and hydrogen, and can be employed directly for industrial or domestic purposes.

Industrial biomass

Some crops are grown specifically for fuel. These include wood, the largest source, especially poplar, willow and bamboo, but also a wide range of other plant material such as wheat, corn, oats, rye and sugarcane. Crops such as wheat produce a high output of energy per hectare of ground, but compete directly with food production. However, waste material in the form of tree stumps, fallen branches and wood chips is a sustainable, and therefore preferable, use of wood biomass.

Biochemical conversion

This process makes use of bacterial enzymes to break down the biomass in an **anaerobic digester**, which yields a combustible gas that can be used for commercial purposes, while the waste products make good agricultural compost.

Biofuels

So-called 'first-generation' biofuels are derived directly from plants such as sugarcane or corn and fermented to produce **bio-ethanol**, which can be used as a motor-vehicle fuel. 'Second-generation' biofuels, in contrast, are derived from non-food-based sources such as agricultural and human waste (e.g. manure). This is a much more environmentally sound use of biomass, but is not so viable economically.

Algal biomass

Algae can be grown specifically for fuel and can be produced five to ten times faster than crops like wheat. The product is fermented to give various combustible products: ethanol, butanol, methane, biodiesel and hydrogen. This method of obtaining biofuels is obviously preferable to exploiting valuable agricultural land.

Production

Biomass accounts for about 10% of global energy use, of which around two-thirds consists of direct burning for cooking and heating purposes in developing countries. Of the remaining third, amounting to just under 15 exa-joules ($1EJ=10^{18}$ joules), about 12.7 EJ is for industrial energy and 2.16EJ is for transportation (e.g. biodiesel and ethanol). The biggest

country	output, peta-joules (PJ)*
Brazil	2725
USA	2600
India	1250
World total	12,706

A. Biomass

country	capacity**, giga-watts*
Germany	35.5
China	1 8.3
Italy	17.6
Japan	13.6
USA	12

B. Solar

country	capacity** mega-watts* (MW)
USA	3,086
Philippines	1,904
Indonesia	1,197
Mexico	958
Italy	843
New Zealand	628
Iceland	575
World total	10,958

C. Geothermal

country	capacity** mega-watts* (MW)
China	91,412
USA	61,091
Germany	34,250
Spain	22,959
India	20,150
UK	10,531
World total	336,000

D. Wind

country	capacity** giga-watts* (GW)
China	212
Brazil	82.2
USA	79
Canada	76.4
Russia	46
World total	970

E. Hydro

Table 11.1 Renewable energy in main producing countries
Data: A, International Energy Authority, 2013; B, Pure Energies, 2014; C, Geothermal Energy Association, 2010; D, Global Wind Energy Council, 2013; E, World-watch Institute, 2010. Notes: * 1PJ = 10^{15} joules; 1GW = 10^9 watts; 1MW = 10^6 watts; ** = installed capacity, not output.

producers of industrial and transportation fuel from biomass are Brazil, the USA (mostly ethanol) and India (table 11.1A); the European Union is also a large contributor with 23% of the global production.

Environmental considerations
Direct burning of biomass is a major pollutant, releasing large quantities of carbon dioxide, carbon monoxide, and nitrous oxides into the atmosphere, together with various other materials of varying toxicity such as heavy metals, in the same way as coal (*see* chapter 9). Dry wood contains about 50% carbon and its combustion, along with that of coal, is the most important source of the increase in post-industrial carbon emissions. Although

wood is theoretically a renewable resource, as it can be regrown, this only occurs on a very long timescale. Whole-tree harvesting of timber is now regarded as harmful to the long-term health of a forest because of its destructive effects on biodiversity (*see* chapter 12). Carbon capture by a mature standing forest is a more effective way of reducing carbon emissions than carbon replacement via the regrowth of timber.

The most effective and environmentally acceptable use of biomass as a fuel is either in power plants that convert the material into electricity through gas production, or into biofuels, using waste products or crops that can be quickly regrown and do not compete for land with valuable food crops.

Solar power
Solar power has three great advantages over other sources of energy: it is free, abundant and, for all practical purposes, inexhaustible. The sunlight falling on just 1% of the Earth's surface is equivalent to the total energy currently provided by all power sources, and it is the only source of power that could theoretically satisfy all our energy requirements into the foreseeable future.

The solar cell
Solar technology is mainly dependent on the **photovoltaic cell** (PVC), which was at first very expensive to produce, but has since become much cheaper with the development of greatly improved devices. The cell works by converting light energy directly into

electricity, using a **semiconductor**, such as silicon or selenium, connected to an electrode. The technique was first demonstrated in 1883 using selenium coated with gold, but was employed mainly for light detectors until 1959, when the US spacecraft Explorer 6 used solar arrays as a source of power. Although several other types of semiconductor have been deployed, using cadmium-telluride, copper-indium-gallium selenide and metal-organic dyes, some of which are more efficient, silicon is the most widely exploited because of its cheapness and because the potential supplies of it are effectively inexhaustible.

A drawback of the earlier devices was that only a small proportion of the potentially available energy was captured by the cell and turned into usable electricity. Moreover, the PVC only works intermittently – neither at night nor on cloudy days – and is much less efficient at higher latitudes. However, the more serious drawback was the production costs, which in the mid-1990s were estimated at about 14 times those of coal-fired power plants; but improvements since then have greatly reduced this differential.

Despite these disadvantages, because of pressure from environmental campaign groups, by 2002, both Japan and Germany had subsidized the installation of large numbers of rooftop solar systems (Fig. 11.1A), producing economies of scale. In 2004, costs of solar power had improved to the point where they were about four times those of nuclear power and three times those of gas, and the big energy companies began to invest in solar power technology. Many countries now subsidize solar power by paying for the electricity that is

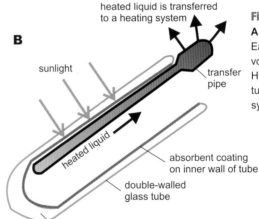

heated liquid is transferred to a heating system

sunlight

transfer pipe

heated liquid

absorbent coating on inner wall of tube

double-walled glass tube

vacuum

Figure 11.1 Solar power
A Solar panels fitted on a house roof. Each panel consists of an array of photovoltaic cells. Image credit: © Shutterstock: Hill120. **B** Cross-section of an evacuated tube collector for a solar space-heating system.

provided by individuals to the national grid by means of a '**feed-in tariff**'.

The main manufacturing countries are Germany, China, Italy, Japan and the USA (table 11.1B). In sunny climates such as those of North Africa or the Middle East, solar power has now become almost competitive with gas.

Solar thermal devices

Solar power can also be used to provide heat from sunlight via a **thermal collector**. These collect heat by absorbing sunlight directly using various types of device. Thermal collectors are particularly useful for providing space heating for residential and commercial buildings. A simple type employs two sheets of metal such as copper, between which is passed the water (or other liquid) to be heated, and would typically be located on the roof of a building. The absorbing plate is coated with dark paint to aid absorption, and may be covered in a glass sheet.

A more sophisticated, but more expensive, method involves a set of double-walled evacuated glass tube collectors, which minimizes heat losses by convection. The heated liquid is conveyed in pipes enclosed within the tubes, attached to the inner wall (Fig. 11.1B). The pipes may be made of plastic or rubber to prevent freeze damage. These devices have the advantage that a greater area of the collector faces the sun.

Bowl-shaped or parabolic collectors may also be employed to concentrate the sunlight into a small area of the collector. However, their much greater cost means that they would only be cost effective on larger buildings.

The simplest type of collector extracts heat directly from the air; in these devices, the air to be heated passes across the absorbing plates. Because of their simplicity and low cost, these are popular for space heating, and are the most efficient of the solar energy technologies.

Geothermal power
Geothermal energy is derived partly from the residual heat created during the formation of the Earth, which amounts now to about 20% of total geothermal energy, and partly from the radioactive decay of minerals such as uranium, which makes up the remaining 80%. This internal heat produces a thermal gradient from the interior of the Earth to the surface, and provides a constant supply of thermal energy, amounting to 44.2 terawatts (44.2 x 10^{12} watts) – about twice the Earth's total energy requirements, at an average heat flow of 0.1 megawatts per square kilometre of the Earth's surface. However, most of this heat cannot be recovered and much of the usable geothermal energy comes from parts of the Earth's crust that experience higher than average heat flow, such as plate boundaries and extensional terrains, where the heat flow can be double that in the plate interiors. Geothermal heat is a valuable resource because it is reliable, almost completely renewable, and when used creates no, or very few, harmful emissions. The only drawback to its employment as a power source is the high capital cost of the power plants, but this is offset by very low (or zero) fuel costs.

History
Hot springs have been used for bathing since Palaeolithic times, and the Romans heated public baths in places such as Bath, in England, from this source. The first commercial geothermal power plant was built in Italy in 1911, and by the 1940s and 1950s hot springs were being exploited in Iceland and New Zealand for domestic heating purposes. At present, Iceland heats over 90% of its homes from geothermal energy, and its capital, Reykjavik, possesses the World's largest district heating system (Fig. 11.2A). In 2010, total world geothermal power capacity amounted to almost 11 gigawatts, with district space heating accounting for a further 28 gigawatts.

The largest producer was the USA, with 29% of the total, followed by the Philippines and Indonesia (table 11.1C).

Technology
In the simplest systems, steam or hot water is piped, or pumped, directly to the user. Vapour-dominated systems from hot-spring areas characterized by geysers (e.g. Fig. 11.2B) are under high pressure, and produce superheated steam at temperatures of up to 300°. This type is typical of the geothermal power plants in New Zealand and Iceland, which are located at, or close to, plate boundaries. However, in most areas of high heat flow, liquid-dominated systems are commoner and can produce temperatures of up to 200°. In these systems, the steam is separated from the hot water, which is returned to the reservoir. The steam drives a turbine, which powers the generator supplying the electricity (Fig. 11.3A). The rate of extraction of heat has to be judged carefully in order not to deplete the heat source; in other words, the system is only renewable if the extraction rate is less than, or equal to, the heat flow into the reservoir.

In lower-temperature liquid-dominated systems, higher efficiency is achieved by passing the hot water through a heat exchanger, in which a liquid with a low boiling point is deployed to carry the heat into the turbine chamber, where it vaporizes (Fig. 11.3B). A condenser then returns the liquid to the exchanger. These systems, termed 'binary-cycle' power plants, require pumps to circulate the hot water and are typical of the geothermal power plants in intra-plate extensional regions such as the western

Figure 11.2 Geothermal power
A Krafla geothermal power station, Iceland. Image credit: © Shutterstock: Tupungato. **B** Geyser, Rotorua, New Zealand.

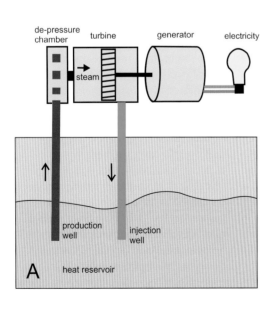
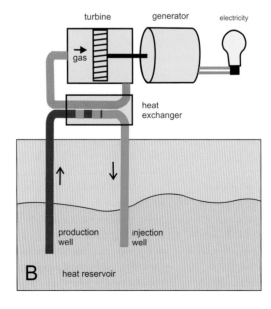

Figure 11.3 Geothermal power systems
Diagrammatic representation of two types of system. **A** The hot water is depressurized into steam, which drives the turbine. **B** The hot water is passed through a heat exchanger that heats a second liquid; this boils at a lower temperature and drives the turbine.

USA and Turkey. Another method of exploiting low-temperature sources is to employ **fracking** techniques to open up hot aquifers at depth (*see* chapter 9). The granites of SW England have been exploited in this way for many years.

Small-scale lower-temperature systems are widely used for district heating or individual domestic heating, and are becoming increasingly popular. These can be deployed almost anywhere merely by exploiting the heat within the bedrock at a relatively shallow depth and amplifying it by employing a **heat pump**, but they are not economic for industrial purposes.

Environmental effects

Hot fluids emanating from the Earth contain natural pollutants such as carbon dioxide, hydrogen sulphide and ammonia, together with some toxic elements such as mercury, and these have to be captured at source. However, the risk of pollution is very much reduced with closed systems using a heat exchanger. In general, however, geothermal power is a cost-effective, reliable and environmentally friendly source of energy.

Wind power

Wind, like solar and geothermal power, is free, abundant and inexhaustible, its main disadvantage being its variability. Because wind power providers do not face fluctuating fuel costs, they can negotiate long-term contracts based only on their known production and upkeep costs. Wind-power technology is also more flexible than conventional power plants since it can be deployed at a variety of scales: single turbines, or groups of turbine units of various sizes

(i.e. wind 'farms'); these can be built relatively rapidly, in contrast with fossil fuel or nuclear plants that may take a decade or more to build. Moreover, even efficient gas-fired power stations need to generate around 100 megawatts to be economical. Single wind turbines can be built near where the power is required, and wind farms can be situated in areas where there is little or no competition from other forms of land use, and which can still be used for agricultural purposes such as cattle grazing.

However, the main reason why wind power technology has proved so popular over the last few decades is because it produces no greenhouse gases, and for that reason has been embraced by many governments as a relatively quick and convenient way of meeting politically-set carbon-emission targets.

History

The first windmill intended to provide electricity was built in 1887 in Scotland by James Blyth, who powered the electric light in his holiday cottage by means of a device with cloth sails. Since then, various types of device have been tried. However, most wind-power now is provided by machines based on a relatively simple design consisting of a horizontal turbine rotor with three long propeller-type blades, connected to a dynamo, and mounted on a tall tower (e.g. Fig. 11.4). In a wind farm, the individual turbines are connected via a low-voltage system to a transformer station and hence to the national grid.

Large wind farms can consist of several hundred turbines, the largest being in China, producing 6 gigawatts, the USA (1.32 gigawatts) and India (1.06 gigawatts). Offshore wind farms

offer the advantages of more frequent and powerful wind, and are less visually obtrusive, but are more expensive to build and maintain. The largest offshore wind farm at present is the London array (630 megawatts), and other large offshore farms are located in Germany and Denmark. By 2014, there were over 200,000 wind turbines worldwide, with a capacity of 336 gigawatts, equivalent to about 4% of total electricity provision. The European Union accounts for over 100 gigawatts of capacity, China for over 90, and the USA for over 60 (table 11.1D).

Disadvantages

Even in the most favourable locations for average wind speed, the turbines can only operate when the wind speed falls between certain limits: too low and the vanes cannot turn, too high and the turbine is shut down to prevent damage. During high-pressure weather systems, wind turbines may not operate for several days at a time. In the most suitable areas, wind turbines only operate at about 45% of their theoretical capacity, and the average is about 33%. Consequently, 'back-up' power generation is required to take over when the wind turbines are not operating. The extent to which this is necessary varies from country to country; recently quoted figures for Scotland showed an average capacity factor of 24% and for Germany, 17.5%. Although improved weather forecasting can help to predict periods of low production, the essential unpredictability of wind power supply inevitably reduces the extent to which it can replace conventional energy sources. It has been estimated that, because of this limitation,

Figure 11.4 Wind power
Wind farm in Aragon, Spain. Image credit: © Shutterstock: Iakov Filimanov.

wind-power will only be able to supply a maximum of about 20% of a region's power requirements. Most European systems at present have wind power proportions varying from less than 5% to 19%. Denmark is the exception at 29% but is able to import spare conventional power from adjoining countries.

Another disadvantage of wind generation systems is their political impact. There are often serious public objections to the building of wind farms in scenically attractive locations (for example, in parts of the Scottish Highlands) on the grounds of their visual impact, coupled with the perceived risk to wildlife, and in some cases to the effects of the noise of the rotors on nearby residents. However, these problems can be overcome to some extent by sensitive siting, and are not so applicable to offshore wind farms.

A more serious disadvantage is the sheer scale of the land taken up by a wind farm; a city of a million households would need about 1000 square miles of wind farm, so when scaled up globally, there is not nearly enough land available to satisfy global power requirements.

Combined systems

The disadvantages of wind power can be offset to some extent by combining it with storage systems, such as hydro-electric pump storage schemes (*see* below), solar power and hydrogen storage. Hydro-electric pump storage uses cheap off-peak wind power to pump water from a downstream reservoir to an upstream reservoir, which can be used again at times of peak demand. However, these are only about 75% efficient and are costly to build. Combination with solar power offers some advantages, since solar power is most efficient during high-pressure weather systems when wind power is least reliable. Probably the best solution for the future lies in the development of more efficient hydrogen fuel cells (*see* chapter 7), which can store wind energy during periods of low demand, and the hydrogen produced would have numerous industrial uses, including motor vehicle fuel.

Economics

The deployment of wind power has been encouraged by governments as a relatively easy and quick way of reaching global greenhouse-gas emissions targets. To this end subsidies are given to landowners in the form of an annual rental income, and in the case of small individual users who connect to the grid from their own homes, of a 'feed-in tariff' based on the excess power that they produce over their domestic requirements. The way that governments subsidize wind power has a big effect on its deployment. Many of the most suitable wind-farm sites are remote from users and incur high transmission costs if the subsidy regime is not set favourably.

Because of declining costs, wind power is now at or near parity economically with conventional sources in Europe, though not yet in the USA. However, the limitations outlined above mean that it is unlikely to satisfy more than about 20% of future power demand.

Hydro-power

Hydro-power makes use either of the kinetic energy of flowing water, or the potential gravitational energy provided by the difference in height between two bodies of water, the higher held in a reservoir behind a dam, and the lower at the outflow below the dam.

History

Small-scale hydro-power schemes have been used by small communities since ancient times, mainly to power corn-grinding mills. Such devices employed a rotating waterwheel part-submerged in flowing water to power the grinding machine. The first electric generator to employ hydro-power was built by William Armstrong in 1878 to provide lighting for his art gallery at Cragside in Northeast England. However, by the end of the nineteenth century, hydro-power stations had been built in many industrial countries – 200 in the USA alone. By the mid-twentieth century several very large schemes had been constructed, including the giant Hoover Dam on the Colorado River at the Nevada–Arizona boundary, south-western USA, which was completed in 1936. In 2010, the USA alone possessed more than 2000 hydro-electric power stations and the total world hydropower capacity was 970 gigawatts (table 11.1E). The Three Gorges Dam in China, with a capacity of 22.5 giga-watts, is currently the World's biggest.

Technology

Conventional hydro-power exploits the gravitational energy of water from rivers and lakes in mountainous regions by creating high-level reservoirs held behind a dam. Water from the reservoir is piped through a turbine into a lower-level outflow. In many of the larger hydro-power stations, the turbine is located inside the dam structure (Fig. 11.5). Some systems consist of a series of pipes from several different reservoirs connected to a downstream power station.

In **pumped storage** systems, there are two reservoirs at different heights. During times of low demand, surplus electricity is used to pump water from the lower reservoir up to the higher, and during times of high demand, the flow is reversed to provide power when needed. This type of system is valuable in creating a store of power and can complement systems with variable output such as wind farms.

Run-of-the-river systems are on a smaller scale, and exploit the natural flow of water, typically in fast-running streams or rivers, and operate either with a small upper reservoir, or without a reservoir. The power has to be used instantaneously, or else the water must be allowed to bypass the turbine. Such installations are particularly useful in providing power for small isolated communities.

Advantages

The principal advantage of conventional hydro-power is its flexibility, and for this reason it is ideal when used as back-up for the highly variable wind power. It is cheap to run, with low maintenance costs, no fuel costs, and there are no harmful emissions. Although expensive to build, the dams and power stations can last for many decades. Some of the earliest plants are still operating up to 100 years after they were built. The reservoirs can also serve a valuable recreational function for leisure activities such as watersports.

Figure 11.5 Hydro-electric power
Diagrammatic representation of a hydro-electric dam and power station.

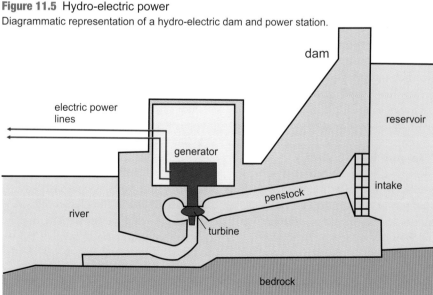

Disadvantages

Land has to be sacrificed to form the reservoirs; in many cases these are located in populated valleys, resulting in the forced expulsion of the inhabitants. There is also an inevitable disturbance to the local ecosystem caused by the increase in flow downstream of the dam, and a corresponding increase in sedimentation in the reservoir. These changes can result in scouring of the river channels and silting up of the reservoirs, with consequential damage to the plant and animal life. Dam failure is rare but, when it occurs, can have serious consequences for the population downstream of the dam. A catastrophic dam burst in Italy in 1963 resulted in about 2000 casualties.

Environmental considerations

Compared with other 'green' energy sources such as solar, wind or nuclear power, hydro-power has many advantages, but there is a limit on how much land can be sacrificed for this purpose. In countries such as Scandinavia and Scotland that have exploited hydro-power for many decades, most of the obvious hydro-power sites have already been used up, and there is not much scope for further development, at least for larger-scale systems.

Tidal and wave power

Compared to conventional hydro-power, the exploitation of wave or tidal power is comparatively recent, and its technology and commercial viability have not, for the most part, been established. However, the possibility of obtaining large amounts of renewable energy from such sources has encouraged several governments to invest in considerable research into the technology.

Tidal power

The natural rise and fall of ocean tides can be exploited by allowing a reservoir to fill up at high tide and letting the water flow out through a turbine at low tide. This type of device is known as a **tidal barrage**. Only a few tidal barrages have been built. The oldest, situated across the Rance Estuary in France, has operated since 1966 and generates 240 megawatts. A barrage has been discussed for the Severn Estuary in England for many years, but the high capital cost has so far prevented the project from being funded.

A tidal barrage consists of a dam containing sluice gates; these are opened to allow water to fill the basin behind the dam during high tide, then closed until low tide, when the water in the basin is allowed to flow out again through turbines located within the sluice gates. There are some environmental concerns over possible changes to the ecosystem and loss of habitat, but the principal objection to the more widespread exploitation of this form of power generation is the very high capital cost.

Tidal currents can also be exploited in places where a narrow channel links two places with different high-tide times. The Pentland Firth, between the Orkney Islands and the Scottish mainland, is such a site, and was approved in 2014, after eleven years of testing, for a tidal power scheme that is projected ultimately to involve 269 turbines mounted on the sea floor with a total capacity of 398 megawatts.

Tidal power, although highly predictable and reliable, only generates power for part of the day, and it will be several years yet before the technology can be adequately evaluated.

Wave power

Various devices designed to exploit wave power are at the experimental stage in several countries, including Scotland, Australia and Portugal, but none are so far operating on a commercial basis. Most of the devices being tested employ floating buoys which use the up and down movement of the waves to operate hydraulic pumps that generate the power.

Renewable energy: future policy

The overwhelming body of scientific evidence supports the conclusion that a significant degree of global warming has taken place since the Industrial Revolution, and that at least part of this warming has resulted from the release of carbon dioxide and other 'greenhouse' gases as a result of human activity (*see* chapter 12). It is also clear that in order to control this increase and avoid potentially catastrophic warming, much of today's exploitation of conventional fuels must be replaced by renewable sources with no, or minimal, greenhouse gas emissions. Unfortunately most of the various energy sources discussed above are not at present economically competitive with fossil fuels, and their deployment must rely on governmental or international intervention in the form of either regulation or subsidy.

To date, only wind power and, to a lesser extent, solar power, have been widely supported by government policy. However, there are limits to the extent to which these can supplant conventional fuels, and it is clear that much more use should be made of the various other alternatives. A balanced policy would exploit geothermal power to a much greater extent, especially in small-scale developments, and this could be coupled with a greater emphasis on energy conservation.

There are limits to the future use of conventional hydro-power, although there is scope for further development of tidal barrages. It is difficult to predict if wave-power has a future until the technology has been adequately tested. Biomass in the form of anaerobic digesters is an obvious candidate for expansion, but biomass use that results in de-forestation, or that competes with agricultural land use, is counter-productive and ought to be discouraged. Finally, many of the problems caused by the intermittent nature of wind and solar power can be solved through the development of more efficient storage technology, such as the use of hydrogen fuel cells.

12 Protecting the Planet

It will have been obvious from the preceding chapters that human interference has caused a number of changes to the environment that, if not corrected or ameliorated, threaten the future viability of the human race. These can be summarized as follows:

1 Over-use and potential exhaustion of mineral resources;
2 Pollution of the environment by toxic materials;
3 Human-induced global warming and climate change;
4 Destruction or degradation of the natural ecology and reduction of biodiversity.

These environmental problems are widely recognized by the scientific community and to a lesser degree by many governments and international agencies, and measures have been put in place to limit some of the damage caused by them.

Mineral resources

The easily accessible reserves of many minerals are being extracted at a rate that is unsustainable, and some will become exhausted at a predictable date in the future unless the rate of extraction is reduced. The rate of use of many resources since the start of industrialization has increased significantly (Fig. 12.1). To some extent this problem is controlled by price: if the cost rises too much, there is an incentive for industry to find an alternative, and in many cases this is a viable solution. However,

there are some metals, regarded as strategic by governments, whose supply is compromised because they are effectively controlled by countries that could, in theory, block their use by competitors. This is a potential source of conflict. Some of the resources that cause concern are noted below.

Metals

The resources of many of the commoner metals listed in chapters 5 and 6 are effectively unlimited, including iron, aluminium, titanium and manganese; others, such as copper, nickel, cobalt and zirconium, have quite large reserves supplemented by considerable recycling of scrap material. However, the existing stocks of a third group have been predicted by some experts to become

exhausted during the present century; these include lead, zinc, molybdenum, tungsten and silver (although about half the lead and much of the silver currently used are recycled).

Both molybdenum and tungsten, along with tantalum, are regarded as **strategic minerals** whose supply is carefully monitored by the leading industrial nations. Of the platinum group metals, both iridium and palladium are regarded as strategic, although much of the latter is recycled. Some of the rare earth metals also have important industrial applications and, because of their rarity and the few sources of supply, the group as a whole is considered to be of strategic importance. Resources of the alkali metals and the alkaline earths are not in danger.

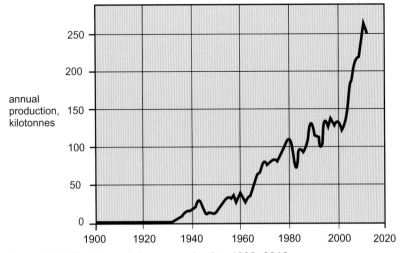

Figure 12.1 World molybdenum production 1900–2010
Annual molybdenum production increases rapidly from about 1930 to the present.

Table 12.1 gives a list of those minerals whose supply is considered by the British Geological Survey (BGS) to be most at risk, together with their leading producing countries. Not all of the listed minerals could be considered to be of strategic importance, and their position on the list depends on a number of factors, including the perceived stability of supply from the major suppliers, as well as on their rarity.

Non-metals

Of the non-metallic minerals, antimony, tellurium and helium all cause concern. Antimony is top of the BGS risk list, with known reserves expected to be exhausted in a few years. The use of tellurium in solar cells is expected to expand in the future, which might endanger resources of that element, while helium supply could be restricted if natural gas extraction is reduced.

Rocks

Several of the rock materials discussed in chapter 8 are of some concern. Cement production, which is vital for all industrialized countries, is dependent on limestone quarrying, which is restricted in many of these countries because of environmental concerns; likewise the extraction of sand and gravel is often limited in highly populated areas. Thus, although deposits of these materials are plentiful, their continued extraction at the present rate may create problems in the future.

Supplies of natural bitumen are potentially restricted, but since much of the bitumen used can be recycled as road surfaces are repaired or replaced, and since natural bitumen can also be sourced from crude petroleum, there is unlikely to be a supply problem.

Environmental pollution

The atmosphere

Pollution of the atmosphere by human activity increased rapidly during the Industrial Revolution, and although we are now much more aware of the damage that is being caused, the cumulative effects are serious, and emissions of certain pollutants are continuing at a dangerous rate. The most damaging pollutants are carbon dioxide, methane, nitrous oxide, sulphur dioxide, halocarbons (e.g. chlorofluorocarbons, or CFCs) and particulates. Burning of biomass, caused for instance by natural forest fires or deliberate forest clearance, is a significant source of atmospheric pollution. According to a World Health Organization report of 2014, air pollution is responsible for the deaths of about seven million people annually.

Carbon dioxide is the most serious pollutant because of its effect as a greenhouse gas (*see* below). However, unlike the others, it is not toxic. Carbon monoxide, in contrast, produced by incomplete combustion of fuel, is highly toxic but is not an important component of the atmosphere. Methane accounts for about 9% of greenhouse gas emissions by the USA, of which 29% comes from natural gas (mainly from leakages) and nearly half from agricultural processes (livestock flatulence and manure) plus landfill sites. Although natural processes use up some atmospheric methane, it is advisable to limit methane production as far as possible because of its effectiveness as a greenhouse gas.

Both sulphur dioxide and nitrous oxide are produced by the burning of fossil fuels in power plants and factories. Sulphur dioxide combines with atmospheric water to form acid rain, which damages the ecology of freshwater lakes and rivers. Nitrous oxide is responsible for the toxic brown haze

Element or element group	symbol	risk index	leading producer
antimony	Sb	8.5	China
platinum group	PGE	8.5	South Africa
mercury	Hg	8.5	China
tungsten	W	8.5	China
rare earths	REE	8.0	China
niobium	Nb	8.0	Brazil
strontium	Sr	7.5	China
bismuth	Bi	7.0	China
thorium	Th	7.0	India
bromine	Br	7.0	USA
carbon (graphite)	C	7.0	China
rhenium	Re	6.5	Chile
iodine	I	6.5	Chile
indium	In	6.5	China
germanium	Ge	6.5	China
beryllium	Be	6.5	USA
helium	He	6.5	USA
molybdenum	Mo	6.0	China
tin	Sn	6.0	China
arsenic	As	6.0	China
silver	Ag	6.0	Peru
tantalum	Ta	6.0	Rwanda

Table 12.1 The 22 most at-risk minerals
List of 22 elements or element groups whose supply is considered most at risk by the British Geological Survey, 2011. The risk index is based on several factors: abundance (scarcity); distribution of reserves, concentration of production, and political stability. Note that a mineral may be widely distributed, but that its production may be concentrated in a few, or even a single, country.

that blankets many cities. In combination with volatile organic compounds such as methane, it reacts with ultraviolet light to create ground-level ozone, which is particularly dangerous.

Halocarbons are now strictly controlled, since it was realized that they were causing damage to the ozone layer.

Particulates are the microscopic particles released by the burning of (mainly) diesel fuel and held in suspension in the air; as well as carbon, these particles include toxic metals such as lead, mercury and cadmium. They are a serious health hazard, responsible for lung and heart disease and for some cancers. These particulates, along with sulphur dioxide and nitrous oxide, create the industrial fog known as **smog**. Until the middle of the twentieth century, smog was common in many British cities such as London, Manchester and Glasgow, but since the use of coal for domestic heating was replaced by gas and electricity, most northern European cities are smog free.

Air quality in Europe and North America is now strictly regulated[1]; however, many world cities are still severely affected by air pollution; of the nine most polluted cities, four are in India and three in China – mainly because of the extensive use of coal.

Many of the problems associated with air pollution are being addressed by the improved design of diesel and petrol engines and by the installation of various devices to capture pollutants from factories and power plant chimneys before being released. Measures primarily designed to reduce

1 [Greater London has been threatened with a large fine by the European Union for breaching air quality standards.]

the release of greenhouse gases will also have a beneficial effect on air quality.

Dangerous atmospheric pollution is also caused by the accidental release of toxic chemicals. Two notorious examples of this are the 1984 Bhopal disaster and the 1986 Chernobyl nuclear explosion. The Bhopal accident, caused by the release of methyl isocyanate gas from a pesticide plant, killed an estimated 4000 people and caused serious permanent injuries to many others. The Chernobyl accident, discussed in chapter 9, killed fewer people immediately, but was responsible for an unknown amount of long-term illness from cancer.

The land

Contamination of the land surface occurs in a variety of ways. Old mine workings and factory sites are a common source of pollution, although modern environmental regulations in the more advanced industrial countries generally prevent more serious cases arising now. Contamination of soils and waterways is frequently caused by accidental or illegal spills of various pollutants including hydrocarbons (e.g. fuel), agricultural herbicides or pesticides, and heavy metals such as lead, cadmium, chromium, zinc and arsenic. Rivers, even in the more strictly regulated countries, are frequently subjected to accidental discharges of industrial waste, untreated sewage and agricultural fertilizers. Illegal mining of gold in a number of developing nations is still a common source of mercury and cyanide pollution.

Radioactive contamination is a serious problem in those countries with nuclear power plants or nuclear armaments because of either accidental

discharge or deliberate disposal of nuclear waste containing material with long-lived radioactivity (*see* chapter 9). Unwise disposal of nuclear waste has produced sites that will remain dangerously radioactive for thousands of years. Discussions over the most effective disposal methods of nuclear waste in the United Kingdom have dragged on for decades without any sign of resolution, but other countries such as Canada and Sweden have decided on deep burial in suitable rock repositories.

The problem of radioactive waste disposal, and the possibility of accidents involving radioactive materials, have led to a vigorous debate, and there is a significant body of informed opinion opposed to the continued deployment of nuclear power, especially in the European Union, where Germany has decided to close down all its nuclear power stations, and several other countries have taken a decision not to build any.

The oceans

Contaminants in soil and rivers eventually end up in the oceans, where some toxic chemicals are ingested by marine organisms, and progressively concentrated through the food chain until they become dangerously magnified in top predators such as seals and whales.

Litter washed into the sea from the shore or dumped from ships is a particular problem. Plastic material is especially hazardous for marine life. Although it is gradually broken down into tiny particles, it never disappears, but is swept up by the oceanic current system and concentrated in certain areas such as the 'Great Pacific Garbage Patch'. This is a vast area – two widely different estimates of its size being 700,000 and 15 million square kilometres respectively – located within the north Pacific. Midway Atoll, situated within this region, receives 20 tons of plastic and other debris every year and the majority of the local albatross population have some plastic material within their digestive systems. Other concentrations are located in the Indian and Atlantic oceans.

Global warming and climate change
The problem
It is accepted by most climate scientists that the human race is facing a serious problem: that the Earth is warming, resulting in increasing climate instability; and that carbon dioxide emissions are at least partly responsible. Average surface temperature has risen by about 0.7°C during the twentieth century, but this has not affected all parts of the Earth equally – the poles have warmed much more, leading, for example, to the reduction of summer sea ice in the Arctic and to the retreat of many glaciers. Several recent years have been the warmest on record. This temperature rise has had a disproportionate effect on the oceans because the extra energy concentrated in the surface layer leads to more extreme climatic effects.

The greenhouse gases
The main greenhouse gases (carbon dioxide and methane) trap heat energy within the atmosphere, and although they are only present in proportions of less than 1%, without them, surface temperature would be around –19°C. Atmospheric carbon has increased from 280ppm by volume before the Industrial Revolution to over 400ppm now. Human-generated emissions of carbon dioxide, due mainly to the burning of fossil fuels, amount to about 28 billion tonnes per year, and if continued at the present rate would double atmospheric carbon dioxide levels from 0.03% to 0.06% within the present century.

The relationship between atmospheric carbon dioxide levels and temperature in the (geologically) recent past can be studied in the Vostok ice core record from the Antarctic ice cap (Fig. 12.2). This clearly shows that temperature and carbon dioxide levels are closely linked over the past 400,000 years; carbon dioxide peaks correspond with inter-glacial periods, and troughs with glacial maxima. But this linkage does not distinguish between cause and effect.

The geological evidence
Global temperatures have always varied: mean temperatures in the early Archaean (c.3.0Ga ago) were probably between 25°C and 35°C (compared to c.14°C now) despite the fact that the Sun was about 25% cooler then. These high early temperatures are explained by the fact that the early atmosphere was rich in carbon dioxide and methane; oxygen only appeared in sufficient quantities to sustain complex animal life about 600Ma ago.

The long-term record from the beginning of the Cambrian Period (540Ma ago) shows that we are presently in an inter-glacial episode within a long-term glaciation that started about 40Ma ago (see Fig. 10.4). Before that, the Earth was much warmer – during the Cretaceous Period, when dinosaurs roamed the continents, deciduous forests reached to the poles. But sea level then was around 200 metres higher than it is now, and much of the low-lying continental area was covered by shallow seas (Fig. 12.3).

The probable geological reasons for these global temperature changes are

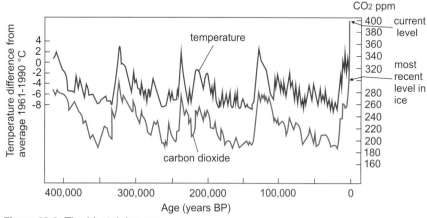

Figure 12.2 The Vostok ice core
Variation in temperature and carbon dioxide levels over the last 100,000 years recorded in the ice core taken at the Vostok research station in Antarctica. Note the close correspondence between the temperature and carbon dioxide curves, and that the CO_2 peaks are all about 280ppm, whereas the current level is over 390ppm.

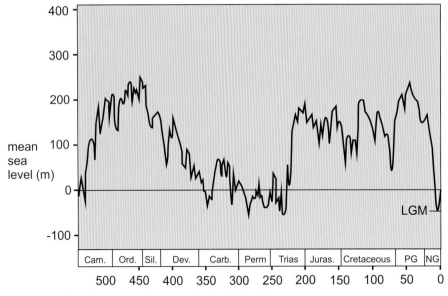

Figure 12.3 Sea-level variation over the last 500 million years
Curve of sea-level variation measured in metres relative to the present level, simplified from the Exxon Sea-Level Curve. LGM, last glacial maximum; NG, Neogene; PG, Paleogene; *see* Appendix, Table A1 for geological time periods.

discussed in chapter 10. However, Earth has always recovered from periods of excessive heat or cold in the past – it has never frozen over completely nor become too hot to sustain life – but the present rate of change is too rapid for the Earth's natural recovery processes to work, so we are in unknown territory. Thus, even if the increase in carbon dioxide does *not* cause global warming, or is not the main cause, it is surely prudent to reduce it to the levels that the Earth has been accustomed to.

The response

Most governments have accepted that carbon emissions have to be reduced. The international Kyoto agreement in 1997 committed 160 countries to an overall reduction in carbon emissions of 5% over 1997 levels with the aim of keeping the projected temperature increase to 2°C. However, this agreement only lasted to 2012 and its extension has not yet been agreed by the USA, China and India; moreover, Canada has withdrawn from it.

Demand for energy will rise as the world population increases and becomes wealthier. Progress towards carbon emission reduction will depend on several factors:

1 Reduction of wasteful energy use;
2 A move away from coal towards gas; coal can be refined into syngas;
3 Installation of carbon capture and storage facilities to reduce the harmful emissions from fossil fuel combustion;
4 Increase in the development and deployment of renewable energy, the various sources of which are

described in chapter 11. To a large extent this is cost-driven and can be encouraged by intelligent use of financial incentives through subsidies and taxes. Much of the criticism of renewable energy, especially wind-power, is based on its variability of supply. However, this can be solved by better techniques of energy storage, especially the making and storing of hydrogen during times of excess supply.

Clearly, some warming has already taken place and cannot be reversed, and more is inevitable even if the planned reductions in carbon emissions are implemented. Meanwhile, action needs to be taken to mitigate the more serious effects such as the drowning of low-lying regions by rising sea level.

Ecological destruction

Human interference in the natural ecology has led to several serious problems, the more critical of which result from habitat destruction and reduction in biodiversity. Habitat destruction has been taking place since forests were first cleared for agricultural purposes in the Neolithic and continues today at an alarming rate. According to the National Institute of Space Research of Brazil, nearly 20% of the Amazon rainforest has disappeared since 1970.

Biodiversity is important because it increases the stability of ecosystems, since diverse communities are more productive and have a greater resilience to ecological disturbance. It is important for providing new drugs derived from biological sources. Yet it is estimated that, because of our actions,

30% of all species will have become extinct by 2050; this is several hundred times the natural rate of species extinction seen in the geological record.

The World Wildlife Fund (WWF) estimated in 2014 that about half of Earth's biodiversity has been lost since 1970. There are many reasons for this; habitat destruction, over-exploitation, and pollution are some of the more important. The widespread practice of monoculture farming has the effect of reducing biodiversity because of the use of pesticides and herbicides. Invasive exotic species in several countries have been so successful (e.g. rabbits in Australia) that the indigenous fauna and flora have been unable to compete. The WWF has compiled a list of the World's most endangered species; 16 of these are on their critically endangered list and include three species of gorilla, two of rhinoceros and two of tiger. Although animals such as these are particularly valued by the public, the disappearance of many others that are relatively unknown is just as serious.

The response

There is widespread public concern over ecological destruction and loss of biodiversity and there are a number of pressure groups such as **Greenpeace** that attempt to influence international and national policy on such matters. There are a few international agreements, such as the UN Convention on Biological Diversity and the Convention on International Trade in Endangered Species (CITES) that have established certain rules aimed at protecting the World's species 'bank'. The convention on biological diversity is particularly important in limiting the rights of individuals or companies to 'steal' biological (especially genetic) material from another country, by establishing the sovereign rights of countries to their own biological resources. The European Union has given certain zoos the role of preserving biodiversity by research and breeding programmes aimed at protecting rare or endangered species.

Some governments have attempted to counter the loss of habitat caused by intensive agricultural practices by subsidizing farmers to establish wildlife corridors and unfarmed field borders to encourage wildlife, and by replanting woodland, but this kind of approach needs to become much more widespread, particularly in some developing countries, where population pressure has resulted in over-exploitation of the land leading to loss of vegetation, soil erosion and flooding.

Human activity has led to what is often referred to as the **Holocene Mass Extinction**, which, in terms of the proportion of species lost, rivals the mass extinctions of the geological past. It is too late to recover what has been lost, but if we are to prevent *Homo sapiens* joining the list, remedial measures need to be taken more seriously.

Glossary

Terms in *italics* are defined in the glossary.

A

actinides [63]: group of radioactive elements with atomic nos. 89–103.

actinium (Ac) [46]: highly radioactive *actinide* element, atomic no. 89 (*see* chap. 7).

aggregate [68]: construction material consisting of gravel-sized crushed rock, added to *cement* to make *concrete* (*see* chap. 8).

alabaster [70]: massive, pale-coloured, crystalline variety of *gypsum* used for ornamental purposes (*see* chap. 8).

albedo [86]: reflectivity of Earth's surface; e.g. ice or snow have high albedo and ocean water low (*see* chap. 10).

alkali metals [23]: group of six highly reactive metallic elements that occupy the left side of the Periodic Table and form strong alkalis in water, the best-known being *sodium* and *potassium*.

alkaline earth metals [23]: group of six metals next to the alkali metals in the Periodic Table, whose oxides form alkalis in water, the best-known being *magnesium* and *calcium*.

alluvial (of a sediment) [69]: deposited in a river valley or on a flood plain (*see* chap. 8).

aluminium (Al) [24]: *post-transition metal*, atomic no. 13 (*see* chap. 5).

amethyst [52]: purple-coloured gem-stone variety of *quartz* (*see* chap. 7).

amino-acid [6]: group of organic acids, based on *nitrogen* and *hydrogen*, forming the basis of proteins.

ammonia [56]: gaseous compound of *nitrogen* and *hydrogen* (NH_3) (*see* chap. 7).

amphibole [16]: group of hydrous *magnesium-iron-aluminium silicate* minerals.

amphibolite [14]: metamorphic rock consisting mainly of *amphiboles* (e.g. *hornblende*) and *plagioclase feldspar*.

anaerobic digester [90]: industrial process in which bacteria are used to break down *biomass* to yield gas fuel plus compostable waste (*see* chap. 11).

andesite [14]: fine-grained intermediate igneous rock (*see* Appendix A2).

anhydrite [47, 70]: *calcium sulphate* mineral ($CaSO_4$) (*see* chap. 6).

anthracite [74]: pure form of coal with a high *carbon* and low volatiles content (*see* chap. 9).

anticyclone [81]: high-pressure weather system defined by winds rotating outwards from the centre, clockwise in the northern hemisphere, anticlockwise in the southern (*see* chap. 10).

antimony (Sb) [54]: *metalloid* element, atomic no. 51 (*see* chap. 7).

apatite (mineral) [58]: complex *oxide* of *calcium*, *fluorine* and *phosphorus* (*see* chap. 7).

Archaean (Eon) [10]: subdivision of geological time (4.0–2.5Ga) (*see* Appendix, Table A2).

argentite [42]: *silver sulphide* (Ag_2S) mineral (*see* chap. 6).

argon (Ar) [62]: *noble gas* element, atomic no. 18 (*see* chap. 7).

arsenic (As) [53]: *metalloid* element, atomic no. 33 (*see* chap. 7).

arsenopyrite [54]: mineral: *sulphide* of *iron* and *arsenic* (FeAsS) (*see* chap. 7).

asphalt: *see bitumen*.

astatine (At) [62]: rare radioactive *halogen* element, atomic no. 85 (*see* chap. 7).

asteroid [4]: small kilometre-sized bodies within the Solar System, especially concentrated in a belt between the orbits of Mars and Jupiter (*see* chap. 2).

asthenosphere [11]: weak layer of the Earth, beneath the *lithosphere*, capable of solid-state flow, over which the *plates* move (*see* Fig. 3.1).

atmosphere [8]: outer gaseous layer of the Earth (*see* Fig. 2.5B).

aurora (australis, borealis) [8, 82]: phenomenon consisting of shimmering coloured lights in the night sky at high southern or northern latitudes respectively, originating in the *ionosphere* (*see* chap. 2).

axinite [52]: complex hydrous *alumino-silicate* of *boron* and *calcium*, with varying amounts of *iron* and *manganese* (*see* chap. 7).

azurite [27]: hydrated *copper carbonate* mineral with a bright blue colour (*see* chap. 5).

B

back-arc basin [18]: extensional basin formed on the upper *plate* of a *subduction* zone (*see* chap. 3).

banded iron formation [10]: rock formed of alternating layers of (mainly) *iron oxide* and *silica*, especially important in the early Precambrian (*see* chap. 2).

barite (or barites) [30]: *barium sulphate* mineral ($BaSO_4$) (*see* chap. 6).

barium (Ba) [47]: *alkaline earth metal*, atomic no. 56 (*see* chap. 6).

basalt [7]: basic igneous rock (*see* Appendix, Fig. A2).

bastnäsite [49]: *carbonate* mineral containing *cerium*, *lanthanum*, *yttrium* and *fluorine* (*see* chap. 6).

bauxite [21]: ore composed of hydrated *aluminium oxides* formed by surface weathering of *alumino-silicates* in tropical climates (*see* chap. 5).

becquerel [63]: unit of measurement of radioactivity = 0.7×10^{-15} grams per cubic metre.

bentonite [68]: type of *clay* consisting mostly of *montmorillonite*, formed, e.g., by weathering of volcanic ash (*see* chap. 8).

bertrandite [46]: hydrous *beryllium silicate* mineral (*see* chap. 6).

beryl [46]: *beryllium-aluminium* silicate mineral, a gemstone; green-coloured varieties are known as *emeralds* (*see* chaps 5, 6).

beryllium (Be) [46]: *alkaline earth* element, atomic no. 4 (*see* chap. 6).

biodiversity [2, 85]: the amount of variability in the Earth's fauna and flora (i.e. numbers and range of different species (*see* chap. 10).

bio-ethanol [90]: gas fuel consisting of *ethanol* produced from plants (*see* chap. 11).

biogas [77]: natural gas composed of *methane* produced from decaying organic matter (*see* chap. 9).

biomass [85]: totality of living organisms; also biological material (e.g. wood chips, organic waste) used as fuel (*see* chap. 10).

biosphere [85]: global system of living organisms (*see* chap. 10).

bismite [36]: *bismuth* tri-*oxide* mineral (Bi_2O_3) (*see* chap. 5).

bismuth (Bi) [36]: *post-transition metal*, atomic no. 83 (*see* chap. 5).

bismuthinite [36]: *bismuth sulphide* mineral (Bi_2S_3) (*see* chap. 5).

bismutite [36]: hydrous *bismuth carbonate* mineral (*see* chap. 5).

bitumen [70]: viscous or semi-solid component of *petroleum*; otherwise known as *asphalt* or *tar* (*see* chap. 8).

bituminous coal [74]: type of coal with a significant volatile component between *lignite* and *anthracite* in purity (*see* chap. 9).

black smoker [18]: underwater active volcanic vent located, e.g., on ocean ridges.

bleach [60]: compound of *chlorine* used as a disinfectant (*see* chap. 7).

bog-iron ore [21]: *iron hydroxide* (*limonite*) formed by acid groundwater extracting iron from the soil in organic marshy conditions (*see* Fig. 4.2C, D).

borax [52]: mineral: complex hydrated *oxide* of *sodium* and *boron* (*see* chap. 7).

boron (Bo) [52]: *metalloid* element, atomic no. 5 (*see* chap. 7).

brass [26]: alloy of *copper* and *zinc* (*see* chap. 5).

Brazil current [85]: Atlantic Ocean current transferring warmer water from the tropics to more southerly latitudes along the east coast of S. America (*see* chap. 10, Fig. 10.1).

breccia [7]: rock consisting of large angular fragments.

brine [16]: water containing dissolved salts, mainly *sodium chloride* (e.g. sea water).

bromine (Br) [61]: *halogen* element, liquid at standard temperature, atomic no. 35 (*see* chap. 7).

bronze [26]: alloy of *copper* and *tin* (*see* chap. 5).

brucite [47]: *magnesium hydroxide* mineral ($Mg(OH)_2$) (*see* chap. 6).

bullion [40]: store of *gold* (or *silver*) used as currency reserve or investment (*see* chap. 6).

butane [76]: type of refined natural gas used as fuel (*see* chap. 9).

C

cadmium (Cd) [36]: *transition metal*, atomic no. 48 (*see* chap. 5).

caesium (Cs) [45]: *alkali metal*, atomic no. 55 (*see* chap. 6).

calcium (Ca) [47]: *alkaline earth metal*, atomic no. 20 (*see* chap. 6).

caliche [61]: *evaporite* deposit containing *oxides* of *sodium* and *iodine* (*see* chap. 7).

californium (Cf) [64]: artificially produced radioactive *actinide* element, atomic no. 98 (*see* chap. 7).

calomel [37]: *mercury chloride* mineral (Hg_2Cl_2) (*see* chap. 5).

cap rock [75]: impermeable rock layer sealing a *petroleum* reservoir (*see* chap. 9, Fig. 9.1).

carbon (C) [55]: *carbon group* element, atomic no. 6 (*see* chap. 7).

carbon group (elements) [55]: group of elements to the right of the *metalloids* in the Periodic Table, of which carbon is the first (*see* chap. 7).

carbonate [55]: negative ion consisting of carbon and oxygen (CO_3); also rock consisting mainly of *calcium* or *magnesium carbonate* (e.g. *limestone, dolomite*).

carbonatite [35]: igneous rock composed mainly of *carbonates* of e.g. *calcium, magnesium, iron, manganese, potassium,*

and also rare elements such as *tantalum* (*see* chap. 5).

carnallite [47]: complex hydrous *potassium-magnesium chloride* mineral (*see* chap. 6).

cassiterite [31]: *tin* oxide mineral (SnO_2) (*see* chap. 5).

celestine [47]: *strontium sulphate* mineral ($SrSO_4$) (*see* chap. 6).

cement [47]: building material (*calcium hydroxide*) obtained by adding water to *lime* (*see* chap. 6).

cerite [49]: *cerium silicate* mineral (*see* chap. 6).

cerium (Ce) [49]: *rare earth* metal, atomic no. 58 (*see* chap. 6).

cerussite [30]: *lead carbonate* mineral ($PbCO_3$) (*see* chap. 5).

CFC [60]: organic *chloro-fluoro-carbon* compound, restricted because of damage to the *ozone layer* (*see* chap. 7).

chain reaction [78]: a series of nuclear reactions, each of which releases neutrons that in turn produce further reactions, and if uncontrolled produces a nuclear explosion (*see* chap. 9).

chalcopyrite [27]: *copper-iron sulphide* mineral ($CuFeS_2$) (*see* chap. 5).

charcoal [55]: carbon-rich fuel made by burning wood under reducing conditions (*see* chap. 11).

chert [13]: rock composed of microcrystalline *silica* (SiO_2) formed e.g. on the deep-ocean floor from micro-fossil deposits.

china clay [16]: pure form of clay formed from *hydrothermally* altered *granite* (*see* chap. 3).

chlorargyrite [42]: *silver chloride* mineral (AgCl) (*see* chap. 6).

chlorine (Cl) [60]: *halogen* element, atomic no.17; a gas at standard temperature (*see* chap. 7).

chlorite [16]: group of hydrous *magnesium-iron-aluminium-silicate* minerals with a flaky appearance formed from *hydrothermal* alteration of (e.g. *pyroxene*) in basic igneous rocks (*see* chap. 3).

chlorophyll [46]: green organic compound, in which *magnesium* is an essential component, used by plants to absorb energy from sunlight (*see* chap. 6).

chondrite [6]: 'stony' *meteorite*,

composed largely of *silicate* minerals, *iron* and *nickel* (*see* chap. 2).

chondrule [6]: small spherical globule of frozen *silicate* melt within a *chondrite* (*see* chap. 2).

chromite [25]: *iron-chromium oxide* (iron chromate) mineral ($FeCr_2O_4$) (*see* chap. 5).

chromium (Cr) [33]: *transition metal*, atomic no. 24 (*see* chap. 5).

cinnabar [37]: *mercury sulphide* mineral (HgS) (*see* chap. 5).

cladding [65]: outer covering of a building with e.g. stone slabs, *cement* etc. (*see* chap. 8).

clast [17]: rock fragment, e.g. within a sedimentary deposit (*see* chap. 3).

cobalt (Co) [27]: *transition metal*, atomic no. 27 (*see* chap. 5).

coke [55]: form of *carbon*, used in *steel* making, produced by heating coal to drive off the gas content (*see* chap. 7).

colemanite [52]: mineral: hydrated *oxide* of *calcium* and *boron* (*see* chap. 7).

coltan [35]: strategically important ore mineral consisting of a mixture of *tantalite* and *columbite* (*see* chap. 5).

columbite (mineral) [35]: *oxide* of *iron*, *manganese* and *niobium* (*see* chap. 5).

concrete [68]: construction material made by mixing *cement*, sand and *aggregate* with water; this solidifies into a rock-like material with a matrix similar to *limestone* (*see* chap. 8).

copper (Cu) [25]: *transition metal*, atomic no. 27 (*see* chap. 5).

core [7]: innermost layer of the Earth, from c.2900km depth to the centre (*see* Fig. 2.5A).

coriolis effect [82]: latitudinal component of movement exerted on wind and ocean current directions because of the Earth's rotation (*see* chap. 10).

corundum [25]: *aluminium oxide* (Al_2O_3), an extremely hard mineral, when combined with small amounts of other elements forms coloured gemstones: e.g. *ruby*, *sapphire* and *emerald* (*see* chap. 5).

crocoite [34]: *lead chromate* mineral ($PbCrO_4$) (*see* chap. 5).

crude oil [76]: *petroleum* recovered directly from its source, before refining (*see* chap. 9).

crust [7]: outermost solid layer of the Earth, varying in thickness from c.7km under the oceans to c.33km, on average, on land (*see* Fig. 2.5A).

crustal assimilation [16]: process where a magma melts and incorporates part of its host rock with a lower melting temperature than the magma (*see* chap. 3).

cryolite [60]: *sodium-aluminium fluoride* mineral (Na_3CaF_6) (*see* chap. 7).

cumulate [15]: early-formed accumulation of minerals with high melting points from a crystallizing magma (*see* Fig. 3.4).

cuprite [27]: *copper oxide* mineral (Cu_2O) (*see* chap. 5).

cyclone [81]: low-pressure weather system defined by winds rotating inwards towards the centre, anticlockwise in the northern hemisphere, clockwise in the southern; extreme examples termed hurricanes or typhoons (*see* chap. 10).

D

degassing [16]: metamorphic process where volatile components are expelled from a rock. (*see* chap. 3).

dehydration [16]: metamorphic process in which water is removed from a rock, typically by heating (*see* chap. 3).

diesel (oil) [76]: type of refined liquid *petroleum*, heavier than petrol, used as motor fuel (*see* chap. 10).

desert rose [70]: crystals of *gypsum* mixed with sand grains, formed in a desert environment (*see* Fig. 8.1E).

diamond (mineral) [56]: gemstone form of *carbon*, produced under very high pressure (*see* chap. 7).

diorite [19]: coarse-grained, intermediate, igneous rock (*see* Appendix, Fig. A2).

dolomite [47]: mineral consisting of a mixture of *calcium*, *magnesium* and *iron carbonate*; also rock (more correctly, 'dolostone') composed largely of dolomite.

dunite [15]: ultrabasic igneous rock composed of *olivine*, with minor *pyroxene* and *spinel*.

dysprosium (Dy) [50]: *rare earth* metal, atomic no. 66 (*see* chap. 6).

E

eclogite [14]: metamorphic rock consisting mainly of *pyroxene* and *garnet*, formed under very high pressure and temperature (*see* chap. 3).

electromagnetic energy [73]: *energy* produced by electromagnetic waves, e.g. electricity, solar radiation (*see* chap. 10).

emerald [46]: bright green gemstone consisting of *beryl* with trace amounts of *chromium* (*see* chaps 5, 6).

energy [72]: the capacity of a system to perform *work* (*see* table 10.1).

Epsom salts [46]: hydrated *magnesium sulphate* mineral ($MgSO_4.7H_2O$), used as a laxative (*see* chap. 6).

erbium (Er) [50]: *rare earth* metal, atomic no. 68 (*see* chap. 6).

ethanol (ethyl alcohol) [108]: liquid *hydrocarbon* compound used, e.g. as a motor fuel.

europium (Eu) [50]: *rare earth* metal, atomic no. 63 (*see* chap. 6).

evaporite [70]: sedimentary rock formed by evaporation of sea water.

exosphere [82]: outermost layer of the *atmosphere* where gas is at extremely low density (*see* chaps. 2, 10).

exothermic [73]: chemical reaction producing heat energy (*see* chap. 10).

F

feed-in tariff [92]: payment made by government to a supplier of electricity (e.g. solar or wind power) based on the amount delivered to the national grid (*see* chap. 11).

feldspar [6]: group of rock-forming *alumino-silicate* minerals incorporating *potassium*, *sodium* and *calcium* in various combinations.

flint [52]: type of microcrystalline *silicon dioxide* (silica) (*see* chap. 7).

fluorine (F) [59]: highly reactive *halogen* element, atomic no. 9 (*see* chap. 7).

fluor-apatite (mineral) [60]: *oxide* of *calcium-fluorine* and *phosphorus* (*see* chap. 7).

fluorspar (fluorite) [59]: *calcium fluoride* mineral (CaF_2) (*see* chap. 7).

fluvio-glacial [69]: process involving a combination of glacial and river transport (*see* chap. 8).

force [72]: strictly, the product of a mass and its acceleration (*see* table 10.1).

fracking (hydraulic fracturing) [76]: recovering *petroleum* from shale by fracturing the rock by injecting fluid under pressure (*see* chap. 10).

fractional crystallization [14]: process in a cooling magma where the constituent minerals with the highest melting temperature crystallize first, leaving the magma depleted in these constituents (*see* Fig. 3.4).

francium (Fr) [46]: extremely radioactive *alkali metal*, atomic no. 87 (*see* chap. 6).

fullerene [55]: artificial form of *graphite* made from ultra-thin sheets rolled into tubes – thus *carbon* fibres (*see* chap. 7).

fuller's earth [69]: type of *bentonite clay* used as a cleaning agent (*see* chap. 8).

G

gadolinium (Gd) [50]: *rare earth* metal, atomic no. 64 (*see* chap. 6).

Gaia [89]: theory of a self-regulating Earth, kept habitable by biological control (*see* chap. 9).

galena [30]: *lead sulphide* mineral (PbS) (*see* chap. 5).

gallium (Ga) [38]: *post-transition metal*, atomic no. 31 (*see* chap. 5).

galvanization [31]: process of coating *steel* with *zinc* to protect against corrosion (*see* chap. 5).

garnet [8]: group of *silicate* minerals composed of varying proportions of, mainly, *aluminium, iron, magnesium* and *calcium*; found mainly in metamorphic rocks (*see* chap. 2).

garnierite [27]: hydrated *magnesium-nickel silicate* mineral with variable composition (*see* chap. 5).

gas hydrate [78]: solidified natural gas found as rock deposits or in permafrost (*see* chap. 10).

gasoline [76]: see *petrol*.

germanium (Ge) [53]: *metalloid* element, atomic no. 32 (*see* chap. 7).

gibbsite [24]: *aluminium hydroxide* mineral (Al$_2$(OH)$_6$) chief constituent of *bauxite* ore (*see* chap. 5).

goethite [21]: hydrous *iron oxide* mineral (FeO.OH) containing small amounts of *nickel* (*see* chap. 4).

gold (Au) [40]: *transition metal*, atomic no. 79 (*see* chap. 6).

graphite (mineral) [55]: form of native *carbon* with a sheet structure (*see* chap. 7).

greenhouse gas [87]: gas that absorbs sunlight, e.g. *carbon dioxide, methane, water* (*see* chap. 9).

Greenpeace [105]: international organization promoting environmental welfare (*see* chap. 12).

greenschist (facies) [19]: suite of low-grade metamorphic rocks, typically containing green *chlorite* or *muscovite*.

guano [58]: *phosphate* deposit formed from bird and bat excreta (*see* chap. 7).

Gulf Stream [85]: Atlantic Ocean current transferring warmer water from the tropics to more northerly latitudes (*see* chap. 9, Fig. 9.1).

gypsum [47]: hydrated *calcium sulphate* (CaSO$_4$.2H$_2$O) (*see* chap. 6).

H

haematite [18]: mineral, common red form of *iron* (ferric) *oxide* (Fe$_2$O$_3$), e.g. rust (*see* chap. 5).

hafnium (Hf) [38]: *transition metal*, atomic no. 72 (*see* chap. 5).

halite [44]: *sodium chloride* mineral (NaCl), common salt (*see* chap. 6).

halogens [59]: group of highly reactive elements (e.g. *chlorine, fluorine*) on the right side of the Periodic Table, either gaseous or liquid at standard temperature (*see* chap. 7).

HCFC [60]: organic *hydro-chloro-fluoro-carbon* compound used, e.g. in refrigeration in place of *CFCs* (*see* chap. 7).

heat pump [95]: device that transfers heat between two bodies by means of a circulating fluid that is alternately compressed and expanded; can be used either for heating or cooling (*see* chap. 11).

helium (He) [62]: *noble gas*, atomic no. 2 (*see* chap. 7).

HFC [60]: organic *hydro-fluoro-carbon* compound designed to replace *HCFCs* (*see* chap. 7).

high-field-strength elements (HFSE) [13]: group of *incompatible elements*, including *phosphorus* and *titanium*, with high atomic charges, that tend to

concentrate in late-stage fractions of a crystallizing magma (*see* chap. 3).

holmium (Ho) [50]: *rare earth* metal, atomic no. 67 (*see* chap. 6).

Holocene mass extinction [105]: mass extinction caused by human intervention during the current geological epoch (*see* chap. 12; Appendix, Fig. A1).

hot spot [12]: part of the Earth's *crust* subject to unusually high heat flow and vulcanicity, typically located within *plates* or at ocean ridges (*see* chap. 3).

Humboldt current [85]: Pacific Ocean current transferring colder water from the Antarctic to more northerly latitudes along the west coast of S. America (*see* chap. 9, Fig. 9.1).

hydration [17]: process where the composition of a mineral or rock is altered by the addition of water, thus 'hydrous' (*see* chap. 3).

hydraulic cement [68]: mixture of *lime* and *clay* used in construction (*see* chap. 8).

hydrocarbons [52]: large group of compounds based on various combinations of *hydrogen* and *carbon* that form the basis for organic life (*see* chap. 7).

hydrogen (H) [52]: the first element in the Periodic Table, atomic no.1 (*see* chap. 7).

hydrosphere [8]: Earth's store of surface water (*see* chap. 9).

hydrothermal [16]: (process): involving transfer of hot aqueous fluid (*see* chap. 3).

I

ilmenite [32]: *iron-titanium oxide* mineral (FeO.TiO$_2$) (*see* chap. 5).

immiscible [15]: (of a liquid): where a liquid component of a magma (e.g. metallic *sulphide*) separates out and crystallizes independently.

incompatible elements [13]: elements that do not fit easily into the crystal structures of the common rock-forming silicate minerals and become concentrated in late-stage fractions of a crystallizing magma (*see* chap. 3).

indium (In) [38]: *post-transition metal*, atomic no. 49 (*see* chap. 5).

iodine (I) [61]: *halogen* element, atomic no. 53; liquid at standard temperature (see chap. 7).

ionosphere [8]: layer within the *thermosphere* where sunlight has caused the gas molecules to be electrically charged and giving rise to the *aurora* phenomenon (see chap. 2).

iridium (Ir) [43]: *platinum group metal*, atomic no. 77 (see chap. 6).

iridium anomaly [43]: widespread deposit at the Cretaceous–Paleocene boundary containing unusually large amounts of *iridium*, attributed to fallout from a meteor impact.

iron (Fe) [23]: *transition metal*, atomic no. 26 (see chap. 5).

J

jetstream [82]: narrow ribbon of very strong wind (up to 200mph), which moves weather systems around the Earth at a height of 9–16km above the surface (see chap. 10).

K

kaolinite [68]: *clay* mineral, hydrated *aluminium silicate* (see chap.8).

kaolinization [16]: hydrothermal process in which *feldspar* is converted into clay minerals (e.g. kaolin) (see chap. 3).

kerosene [76]: light fuel oil used, e.g. in jet engines and domestic heating boilers; also known as 'paraffin' (see chap. 9).

kimberlite [56]: type of *peridotite* formed under high pressure and temperature at great depth; source of natural *diamonds* (see chap. 7).

kinetic energy [22]: the *energy* possessed by a moving body (see chap. 9).

komatiite [20]: fine-grained ultrabasic rock, mostly confined to Archaean volcanic deposits (see Appendix, Fig. A2).

krugerrand [20]: gold coin issued by the South African government, used as an investment (see chap. 6).

krypton (Kr) [63]: *noble gas* element, atomic no, 36 (see chap. 7).

L

Labrador Current [85]: Atlantic Ocean current transferring colder water from the Arctic along the east coast of Canada (see chap. 10, Fig. 10.1).

lanthanide(s) [48]: also known as '*rare*

earth' metals: group of rare elements with atomic numbers between 57 and 71 in the Periodic Table (see chap. 6).

lanthanum (La) [49]: *rare earth* metal, atomic no. 57 (see chap. 6).

large-ion-lithophile elements (LILE) [13]: group of *incompatible elements*, including rubidium, potassium and strontium, with large ionic radii, that tend to concentrate in late-stage fractions of a crystallizing magma (see chap. 3).

laterite [21]: type of tropical weathering deposit formed in hot, wet climates and enriched in *iron* and *aluminium hydroxides* (see chap. 4, Fig. 4.2B).

lepidolite [44]: complex *lithium mica* mineral (see chap. 6).

lherzolite [13]: ultrabasic rock thought to (largely) compose the *asthenosphere* and to be the parent rock of *basalt*.

lignite [73]: brown, impure, partly carbonized, coal (see chap. 9).

lime [47, 68]: *calcium oxide* (CaO); compound obtained by heating *limestone* (see chap. 6).

limestone [12]: rock composed of *calcium carbonate*.

limonite [21]: hydrous (ferric) *iron oxide* mineral, with variable composition (see chap. 5).

lithium (Li) [44]: *alkali metal*, atomic no. 3 (see chap. 6).

lithosphere [11]: strong upper layer of the solid Earth from which the tectonic *plates* are formed, consisting of the *crust* and part of the upper *mantle* (see Fig. 3.1).

lutetium (Lu) [51]: *rare earth* metal, atomic no. 71 (see chap. 6).

M

magnesite [47]: *magnesium carbonate* mineral ($MgCO_3$) (see chap. 6).

magnesium (Mg) [46]: *alkaline earth metal*, atomic no. 12 (see chap. 6).

magnetite [23]: (ferric) *iron oxide* mineral (Fe_3O_4) with magnetic properties (see chap. 5).

malachite [27]: hydrated *copper carbonate* mineral with a bright green colour ($CuCO_3 . Cu(OH)_2$ (see chap.5).

manganese (Mn) [33]: *transition metal*, atomic no. 25 (see chap. 5).

mantle [7]: layer of the solid Earth between the *crust* and the *core* (see Fig. 2.5A).

marble [65]: metamorphic rock consisting largely of *calcite* and/or *dolomite*; also a *limestone* used in the building trade (see chap. 8).

mercury (Hg) [37]: *transition metal*, atomic no. 80 (see chap. 5).

mesosiderite [7]: type of brecciated stony-iron meteorite, possibly of collisional origin (see chap. 2).

mesosphere [8]: layer of the *atmosphere* above the *stratosphere*, merging upwards into the *thermosphere* (see Fig. 2.5B).

metalloid(s) [52]: group of elements to the right of the *post-transition metals* in the Periodic Table, the best-known being *silicon* (see chap. 7).

metasomatism [16]: metamorphic process accompanied by fluid transfer (see chap. 3).

mica [17]: group of hydrous *potassium-aluminium silicate* minerals with a flaky appearance; varieties include *iron*, *magnesium* or *lithium* in addition.

milk of magnesia [46]: *magnesium hydroxide* mineral ($Mg(OH)_2$), used as a remedy for indigestion (see chap. 6).

Minamata Convention [37]: UN environmental programme specifying safe limits for mercury use (see chap. 5).

molybdenite [29]: *molybdenum sulphide* mineral (MoS_2) (see chap. 5).

molybdenum (Mo) [29]: *transition metal*, atomic no. 42 (see chap. 5).

monazite [49]: *oxide* mineral containing *cerium, lanthanum, yttrium* and *phosphorus* (see chap. 6).

montmorillonite [68]: *clay* mineral, hydrated *magnesium-aluminium silicate*; main constituent of *bentonite* (see chap. 8).

mortar [68]: building material, made by mixing *lime* and sand (see chap. 8).

muscovite (*mica*) [16]: complex hydrous *potassium-aluminium silicate* mineral (see chap. 3).

N

native (element) [25]: occurring in an uncombined state rather than as a compound, e.g. 'native *silver*', 'native copper', etc.

natural resources [1]: Earth resources that occur naturally, without human intervention (*see* chap.1).

neodymium (Nd) [49]: *rare earth* metal, atomic no. 60 (*see* chap. 6).

neon (Ne) [62]: *noble gas* element, atomic no. 10 (*see* chap. 7).

nickel (Ni) [27]: *transition metal*, atomic no. 18 (*see* chap. 5).

niobium (Nb) [35]: *transition metal*, atomic no. 41 (*see* chap. 5).

nitrogen (N) [56]: *carbon group* element, atomic no.7 (*see* chap. 7).

noble gases [62]: group of non-reactive gaseous elements at the extreme right-hand side of the Periodic Table (*see* chap. 7).

nuclear fission [78]: process of splitting *uranium* atoms into lighter atoms such as *plutonium* by bombardment with neutrons, thus releasing energy (*see* chap. 9).

nuclear fusion [3]: process of combining *hydrogen* atomic nuclei to form *helium* with the release of large amounts of energy (*see* chap. 2).

nucleosynthesis [4]: the process of creating more complex atoms from existing simpler atoms by nuclear reactions (*see* chap. 2).

O

olivine [6]: group of rock-forming *silicate* minerals containing variable combinations of (mainly) *iron, magnesium* and *calcium* (*see* Appendix, table A1).

open-cast (mining) [19]: carried out at the surface, rather than underground.

orogenic belt [13]: mountain belt resulting from collision or subduction of tectonic *plates*.

orthoclase (feldspar) [45]: potassium-aluminium silicate mineral ($KAlSi_3O_8$) (*see* chap. 6).

osmium (Os) [42]: *platinum group* metal, atomic no. 76 (*see* chap. 6).

oxygen (O) [56]: *carbon group* element, atomic no. 8 (*see* chap. 7).

ozone [8]: molecule consisting of three atoms of *oxygen*; the **ozone layer** [82] lies within the *stratosphere* and protects Earth from ultraviolet radiation (*see* Fig. 2.5B).

P

palladium (Pd) [42]: *platinum group* metal, atomic no. 46 (*see* chap. 6).

pallasite [7]: type of stony-iron *meteorite* containing roughly equal proportions of iron-nickel and silicate (*see* chap. 2).

Pangaea [85]: supercontinent consisting of all the Earth's continental landmass that existed during much of the Upper Palaeozoic Period (*see* chap. 10; Appendix, table A1).

panning [22]: technique used by amateur prospectors to separate out heavy minerals such as *gold* from river gravel using a shallow dish or pan.

partial melting [13]: process where part of an igneous rock melts to yield a liquid whose composition corresponds to that of those minerals with lower melting points than the rock as a whole (*see* chap. 3).

particulates [102]: microscopic particles released by the combustion of fossil fuels, especially *diesel* oil, and which are a serious health hazard (*see* chap. 12).

passive margin [12]: continental margin lacking vulcanicity or significant tectonic activity, formerly an *extensional rift* (*see* Fig. 3.2C).

patronite [34]: *vanadium sulphide* mineral (*see* chap. 5).

peat [73]: consolidated deposit of decayed plant material produced under reducing conditions in a bog and used as fuel (*see* chap. 10).

pegmatite [32]: late-stage volatile-rich product of a cooling magma characterized by large crystals and typically forming intrusive veins or dykes.

pentlandite [27]: *nickel-iron sulphide* mineral (*see* chap. 5).

peridotite [8]: *ultrabasic* rock composed mostly of *olivine* and *pyroxene* (*see* Appendix, table A1).

perovskite [8]: *oxide* of *calcium* and *titanium* with a very dense molecular structure (*see* chap. 2).

petalite [44]: *lithium-aluminium silicate* mineral ($AlSi_4O_{10}$) (*see* chap. 6).

petrol: ('gasoline' in N. America) [76]: refined light *petroleum* used, e.g. as motor fuel (*see* chap. 10).

petroleum [70]: mixture of organic *hydrocarbon* compounds forming natural oil or gas (*see* chap. 9).

phosphor bronze [58]: alloy of *copper* and *phosphorus* (*see* chap. 7).

phosphorus (P) [58]: *carbon group* element, atomic no. 15 (*see* chap. 7).

photic zone [84]: uppermost layer of the oceans through which sunlight penetrates (*see* chap. 9).

photosynthesis [10]: process whereby organisms (e.g. plants) consume *carbon dioxide* and expel *oxygen* using solar energy.

photo-voltaic cell (PVC) [91]: device that converts solar energy into electric current using a *semiconductor* (*see* chap. 11).

pitch [70]: type of *bitumen* formerly used, e.g. to make ships' hulls watertight (*see* chap. 8).

pitchblende [48]: *uranium oxide* ore with variable composition, usually containing other elements (e.g. *thorium, zirconium, lead*) (*see* chap. 7).

placer deposit [17]: type of ore deposit formed by the redistribution and concentration of ore by water, e.g. in a river system (*see* chap. 3).

plagioclase [115]: group of rock-forming *alumino-silicate* minerals containing varying proportions of *sodium* and *calcium* (*see* Appendix, table A1).

planetary differentiation [7]: process whereby the constituents of a planetary body are physically separated, e.g. by fractional melting, fractional crystallization or evaporation.

planetesimals [3]: small kilometre-sized bodies from which the planets of the solar system evolved (*see* chap. 2).

plaster of Paris [47]: type of *gypsum* used, e.g. for setting broken bones.

plate tectonics [11]: set of processes involving the relative movements of plates (pieces of the Earth's *lithosphere*), responsible for vulcanicity, mountain building and other geological phenomena, and driven by solid-state flow in the Earth's *mantle*.

platinum (Pt) [42]: *transition metal*, atomic no. 78 (*see* chap. 6).

platinum group (elements) (PGE) [42]: group of rare and strategically important metals: *platinum, iridium, palladium* and *rhodium*, that occur next to *gold* and *silver* in the Periodic Table (*see* chap. 6).

plutonium (Pu) [64]: highly radioactive *actinide* element, atomic no. 94, artificially produced from *uranium* breakdown (*see* chap. 7).

polonium (Po) [55]: highly radioactive *metalloid* element, atomic no. 84 (*see* chap. 7).

pollucite [39]: complex *caesium alumino-silicate* mineral containing trace amounts of *thallium* (*see* chap. 5).

porphyritic [19]: igneous rock characterized by larger crystals within a finer-grained matrix; the term 'porphyry' is typically applied to acid or intermediate varieties.

porphyry copper (ore deposit) [19]: type of deposit typically associated with acid igneous intrusions and volcanic vents above active *subduction* zones (*see* chap. 4, Fig. 4.1A).

post-transition metals [23]: group of metallic elements at the right-hand end of the Periodic Table between the *transition metals* and the *metalloids*, the best-known being *silicon* and *arsenic*.

potash [45]: general term for *potassium* salts, e.g. *chlorides* or *sulphates* (*see* chap. 6).

potassium (K) [45]: *alkali metal*, atomic no. 19 (*see* chap. 6).

potential energy [72]: the gravitational *energy* possessed by a body because of the height through which it can fall (*see* chap. 9).

power [72]: strictly, the rate at which *work* is done, i.e. work divided by time (*see* table 9.1).

praseodymium (Pr) [49]: *rare earth* metal, atomic no. 59 (*see* chap. 6).

producer gas [90]: gas fuel composed mainly of *carbon monoxide* and *hydrogen*, obtained by burning of *biomass* (*see* chap. 11).

promethium (Pm) [49]: *rare earth* metal, atomic no. 61 (*see* chap. 6).

propane [76]: type of refined natural gas used as fuel (*see* chap. 9).

Proterozoic (Eon) [10]: unit of geological time (*see* Appendix, Fig. A1).

pump-storage (hydro-power systems) [97]: hydro-electric power station that pumps water from a lower to a higher reservoir during low electricity demand (*see* chap. 11).

PVC [61]: poly-vinyl *chloride* – organic compound; type of plastic (*see* chap. 7); also, photo-voltaic cell.

pyrargentite [42]: *silver-antimony sulphide* mineral (Ag_3SbS_3) (*see* chap. 6).

pyrite [23]: *iron sulphide* mineral (FeS_2) (*see* chap. 5).

pyrochlore (mineral) [36]: complex *hydroxide* of *tantalum* and *niobium*, with *sodium, calcium* and *fluorine* (*see* chap. 5).

pyrolusite [33]: *manganese dioxide* mineral (MnO_2) (*see* chap. 5).

pyroxene [6]: group of rock-forming *silicate* minerals containing variable combinations of (mainly) *iron, magnesium, calcium* and *aluminium* (*see* Appendix, table A1).

pyrrhotite [18]: complex *iron sulphide* mineral, often containing *nickel* in addition (*see* chap. 4).

Q

quicklime [47]: calcium oxide (see '*lime*') (chap. 6).

R

radium (Ra) [48]: highly radioactive *alkaline earth metal*, atomic no. 88 (*see* chap. 6).

radon (Rn) [63]: radioactive *noble gas* element, atomic no. 86 (*see* chap. 7).

rare earths [48]: see *lanthanides*.

renewable resources [1]: Earth resources that are replaceable, thus effectively unlimited (*see* chap. 1).

reserve [1]: a *resource* calculated to be exploitable economically.

reservoir rock [75]: porous, permeable rock containing natural oil and/or gas (*see* chap. 9, Fig. 9.1).

resource [1]: stock of a substance that is potentially available to be exploited (*see* chap. 1).

rhenium (Re) [38]: *transition metal*, atomic no. 75 (*see* chap. 5).

rhodium (Rh) [43]: *platinum group* metal, atomic no. 45 (*see* chap. 6).

rhodochrosite [33]: *manganese carbonate* mineral ($MnCO_3$) with an attractive pink colour (*see* chap. 5, Fig. 5.4).

rhyolite [14]: fine-grained acid igneous rock consisting typically of *quartz, feldspar* and *mica* (*see* Appendix, Fig. A2).

rift [12]: linear extensional zone of the Earth's *crust* subject to divergent *plate* movements (*see* Fig. 3.2B).

Rodinia [86]: supercontinent consisting of most of the continental landmass that existed during the Late Proterozoic Eon (*see* chap. 10; Appendix A1).

rubidium (Rb) [45]: *alkali metal*, atomic no. 37 (*see* chap. 6).

ruby [25]: red gemstone variety of *corundum* (*see* chap. 5).

run-of-the-river (hydro-power system) [97]: hydro-electric power plant that employs the *kinetic energy* of the flowing water in a river (*see* chap. 11).

ruthenium (Ru) [44]: *platinum group* metal, atomic no. 44 (*see* chap. 6).

rutile [32]: *titanium dioxide* mineral (TiO_2) (*see* chap. 5).

S

saltpetre [54]: potassium nitrate mineral (KNO_3) used in explosives (*see* chap. 6).

samarium (Sm) [49]: *rare earth* metal, atomic no. 62 (*see* chap. 6).

sapphire [25]: blue gemstone variety of *corundum* (*see* chap. 5).

scandium (Sc) [37]: *transition metal*, atomic no. 21 (*see* chap. 5).

scheelite [31]: *calcium-tungsten oxide* mineral ($CaWO_4$) (*see* chap. 5).

SEDEX (ore deposit) [20]: short for 'sedimentary-exhalative': ore deposits that are *stratiform*, i.e. deposited in layers parallel to bedding, and with no obvious local volcanic source (*see* chap. 4).

selenium (Se) [58]: *carbon group* element, atomic no.34 (*see* chap. 7).

semiconductor [92]: device or material (e.g. *silicon, germanium*) intermediate between a conductor and an insulator, whose electrical properties can be controlled by small amounts of another element; fundamental to the electronics industry (*see* chap. 11).

serpentine [16]: group of hydrous *magnesium silicate* minerals formed by *hydrothermal* alteration of ultrabasic igneous rock (e.g. *peridotite*) (*see* chap. 3).

serpentinite [21]: rock composed largely of *serpentine* (*see* chap. 4).

shear zone [16]: planar zone of highly deformed rocks, deep-seated equivalent of a fault.

silicon (Si) [52]: *metalloid* element, most abundant in the Earth's *crust*, atomic no. 14 (*see* chap. 7).

silicate [3]: large group of minerals in which various elements, such as metals and *alkalis* are combined with *silicon dioxide*; the most important rock-forming group (*see* Appendix, table A1).

silver (Ag) [42]: *transition metal*, atomic no. 47 (*see* chap. 6).

skarn [16]: hydrothermally altered host rock (especially *carbonate*) at the margin of an igneous intrusion (*see* chap. 3).

slaked lime [47]: *calcium hydroxide*, made by adding water to *lime* (*see* chap. 8).

slate [67]: metamorphosed mudstone with well-developed cleavage (*see* chap. 8).

smithsonite [31]: *zinc carbonate* mineral ($ZnCO_3$), often containing *cadmium* and other trace elements (*see* chap. 5).

smog [90]: dense industrial fog produced in cities, caused by pollutants, e.g. *carbon dioxide*, *nitrous oxide*, carbon particles etc., caused mainly by coal burning (*see* chap. 11).

Snowball Earth [86]: state of complete or near-complete ice cover over the Earth (*see* chap. 10).

soda ash [44]: hydrated sodium carbonate mineral occurring in evaporite deposits in arid regions.

sodium (Na) [44]: *alkali metal*, atomic no. 11 (*see* chap. 6).

solar nebula [3]: the cloud of gas and dust from which the Solar System evolved (*see* chap. 2).

solder [31]: process where melted *tin* is used to join two other metals.

sorting [17]: process where particles of different sizes (e.g. pebbles, sand and clay) are concentrated in separate groups or deposits (*see* chap. 3).

sperrylite [42]: *platinum arsenide* mineral (PtAs2), containing trace quantities of other *platinum group* metals (*see* chap. 6).

sphalerite [31]: *zinc-iron sulphide* mineral (ZnFeS) (*see* chap. 5).

spinel [25]: group of *oxide* minerals (e.g. *magnetite*) containing varying proportions of *iron*, *magnesium*, *aluminium* and *chromium* (*see* chap. 2).

spodumene [44]: *lithium-aluminium silicate* mineral ($LiAlSi_2O_6$) (*see* chap. 6).

stainless steel [27]: alloy produced by adding *nickel* to *steel* to protect against corrosion (*see* chap. 5).

steel [23]: alloy of *iron*, combined with various other elements (e.g. *carbon*, *vanadium*, *titanium*) to convey desired properties such as high strength (*see* chap. 5).

sterling silver [48]: *silver* alloy, of 93% purity, used in trophies, etc. (*see* chap. 6).

stock [19]: medium-sized intrusive igneous body, typically the feeder to a volcano (*see* chap. 4).

stibnite [54]: *antimony sulphide* mineral (Sb_2S_3) (*see* chap. 7).

stockwork [18]: zone of mineralized veins in the outer part of an igneous body (e.g. a volcanic vent) (*see* chap. 4).

strategic minerals [100]: those considered to be essential to industry and whose supplies are closely monitored (*see* chap. 12).

stratiform (ore deposit) [18]: type of ore deposit formed in a layer parallel to bedding (*see* chap. 4).

stratosphere [8]: layer of the *atmosphere* above the *troposphere* (*see* Fig. 2.5B).

stromatolite [10]: primitive colonial marine organism consisting of successive layers of cyano-bacteria; their fossils are preserved as *carbonate* mounds, particularly in Archaean rocks (*see* chap. 2).

strontianite [47]: strontium carbonate mineral ($SrCO_3$) (*see* chap. 6).

strontium (Sr) [47]: *alkaline earth* metal, atomic no. 38 (*see* chap. 6).

subduction [11]: process where *lithosphere* plates descend into the Earth's *mantle* causing vulcanicity at the surface (*see* Fig. 3.3).

sulphur (S) [58]: *carbon group* element, atomic no. 16 (*see* chap. 7).

supernova [3]: giant type of star in which the heavier elements are created (*see* chap. 2).

syenite [25]: coarse-grained igneous rock similar to granite, but of alkaline affinity, with little or no quartz.

sylvite [39]: *potassium chloride* mineral (KCl) (*see* chap. 5).

syngas [74]: synthetic combustible gas, containing a mixture of *hydrogen* and *carbon monoxide* produced by heating coal under pressure (*see* chap. 9).

T

tantalite (mineral) [35]: *oxide* of *iron*, *manganese* and *tantalum* (*see* chap. 5).

tantalum (Ta) [34]: *transition metal*, atomic no. 73 (*see* chap. 5).

tar [70]: see *bitumen*.

tarmacadam ('tarmac') [71]: is tar mixed with *aggregate* for road surfacing etc. (*see* chap. 8).

technetium (Tc) [39]: rare radioactive *transition metal*, atomic no. 43 (*see* chap. 5).

tellurite [54]: *tellurium oxide* mineral (TeO_2) (*see* chap. 7).

tellurium (Te) [54]: rare *metalloid* element, atomic no. 52 (*see* chap. 7).

terbium (Tb) [50]: *rare earth* metal, atomic no. 65 (*see* chap. 6).

thallium (Tl) [39]: *post-transition metal*, atomic no. 81 (*see* chap. 5).

Theia [9]: hypothetical twin planet of Earth during its early history, thought to have collided with it giving rise to the Moon (*see* chap. 2).

thermal collector [92]: device, usually roof-mounted, to collect solar heat to produce warm air for a building (*see* chap. 11, Fig. 11.1B).

thermal energy [73]: the amount of heat possessed by a body (*see* chap. 9).

thermocline [84]: transitional zone in the oceans between the surface layer and the deep zone, marked by a sharp decrease in temperature and salinity (*see* chap. 10).

thermosphere [8]: layer of the *atmosphere* above the *mesosphere*, merging upwards into the *exosphere* (*see* Fig. 2.5B).

thorium (Th) [64]: radioactive *actinide* element, atomic no. 90 (*see* chap. 7).

thulium (Tm) [51]: *rare earth* metal, atomic no. 69 (*see* chap. 6).

tidal barrage [98]: hydro-electric power

plant situated in a dam ('barrage') across a bay or estuary to exploit the tidal flow (*see* chap. 11).

tin (Sn) [31]: *post-transition metal*, atomic no. 50 (*see* chap. 5).

titanium (Ti) [32]: *transition metal*, atomic no. 22 (*see* chap. 5).

tourmaline [52]: complex *alumino-silicate* of *boron* with various other elements e.g. *iron*, *magnesium* or other *alkali metals* (*see* chap. 7).

transition metals [23]: all the metallic elements in the middle part of the Periodic Table.

troposphere [8]: the lowest part of the *atmosphere*; where weather originates (*see* Fig. 2.5B).

tungsten (W) [31]: *transition metal*, atomic no. 74 (*see* chap. 5).

U

uraninite [48]: complex *uranium oxide* mineral (commonly known as *pitchblende*) (*see* chap. 4).

uranium (U) [54]: radioactive *actinide* element, atomic no. 92 (*see* chap. 7).

urea [58]: organic compound excreted by animals, a source of *phosphorus* (*see* chap. 7).

V

vanadinite [34]: complex *lead-vanadium* oxide mineral, also containing *chlorine* (*see* chap. 5).

vanadium (V) [34]: *transition metal*, atomic no. 23 (*see* chap. 5).

verdigris [27]: coating of green *copper carbonate* on copper exposed to acid rainwater (*see* chap. 5).

W

witherite [48]: *barium carbonate* mineral ($BaCO_3$) (*see* chap. 6).

wolframite [31]: *iron-manganese-tungsten oxide* mineral ($FeMnWO_4$) (*see* chap. 5).

work [72]: strictly, the product of a *force* and the displacement achieved by it (*see* table 9.1).

X

xenolith [13]: inclusion of rock from a different source within an igneous rock.

xenon (Xe) [63]: *noble gas* element, atomic no. 54 (*see* chap. 7).

xenotime [38]: *yttrium phosphate* mineral (*see* chap. 5).

Y

ytterbium (Yb) [51]: *rare earth* metal, atomic no. 70 (*see* chap. 6).

yttrium (Y) [38]: *transition metal*, atomic no. 39 (*see* chap. 5).

Z

Zechstein [45]: the upper part of the Permian period (*see* Appendix, Fig. A1).

zinc (Zn) [30]: *transition metal*, atomic no. 30 (*see* chap. 5).

zircon [32]: mineral consisting of *zirconium silicate*, and containing small quantities of *uranium*; the proportion of radioactive decay products can be used to date the mineral (*see* chap. 2).

zirconium (Zr) [32]: *transition metal*, atomic no. 40 (*see* chap. 5).

Appendix

Geological Time

Geological time is divided into *eons*, *eras*, *periods* and *epochs* (Figure A1). Eons, the first-order divisions, comprise the *Hadean* (from the formation of the Earth to 4Ga), followed by the *Archaean* (4–2.5Ga), the *Proterozoic* (2.5Ga–542Ma) and the *Phanerozoic* (542Ma to Present). The Proterozoic is divided into Early, Mid and Late, and the Phanerozoic into *Palaeozoic*, *Mesozoic* and *Cenozoic* Eras. Each of the eras is further subdivided into periods (e.g. the Cambrian) and the periods into epochs. Only the epochs of the Cenozoic are shown in Figure A1. Note that 1Ga = 1000 million years and 1Ma = 1 million years.

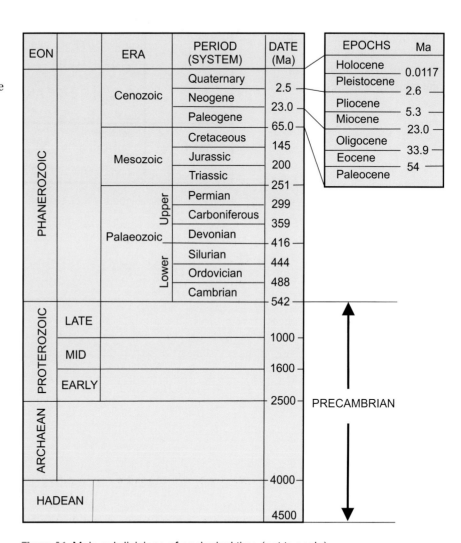

Figure A1 Main subdivisions of geological time (not to scale)

Igneous rock types

Igneous rocks are divided into categories depending on a) their grain size, and b) their proportion of light-coloured minerals, such as quartz or feldspar, compared with dark minerals such as hornblende, pyroxene or olivine. Figure A2 shows the main types of igneous rock together with the more important silicate minerals that they contain.

grain size	acid	intermediate	basic	ultrabasic
coarse	GRANITE	DIORITE	GABBRO	PERIDOTITE
medium	MICRO-GRANITE	MICRO-DIORITE	DOLERITE	
fine	RHYOLITE	ANDESITE	BASALT	KOMATIITE
main minerals	feldspar quartz mica hornblende	feldspar hornblende biotite	feldspar pyroxene olivine	pyroxene olivine

Figure A2 Principal igneous rock types and their main silicate minerals

Selected further reading

British Geological Survey, NERC. Commodities factsheets (online).

British Geological Survey, NERC. Mineral planning factsheets (online).

British Geological Survey, NERC. World mineral production 2005–2009 (online).

Burroughs, W.J., Crowder, B., Robertson, T., Valler-Talbot, V. & Whitaker, R. 1996. *Weather*. Harper-Collins, London.

Evans, A.M., 2011. *Ore geology and industrial minerals, an introduction*. Blackwell Science.

Gill, R. 2010. *Igneous rocks and processes: a practical guide*. Wiley-Blackwell, Chichester, UK.

Gribble, C.D., 1988. *Rutley's elements of mineralogy*. Springer, Netherlands.

Helm, D., 2012. *The carbon crunch: how we're getting climate change wrong – and how to fix it*. Yale University Press, New Haven & London.

Lovelock, J., 2000. *Gaia: a new look at life on Earth*. Oxford University Press.

Mason, J., 2015. *Introducing mineralogy*. Dunedin, Edinburgh.

Park, G., 2006. *Introducing geology: a guide to the world of rocks*. Dunedin, Edinburgh.

Parsons, P. & Dixon, G. *The Periodic Table: a field guide to the elements*. Quercus Science.

Roberts, P., 2004. *The end of oil*. Bloomsbury, London.

Smil, V., 2006. *Energy: a beginners' guide*. Oneworld.

United States Energy Information Administration: International energy statistics (online).

United States Geological Survey 2014. Mineral commodities summaries (online).

Zalasiewicz, J. & Williams, M., 2012. *The Goldilocks planet*. Oxford University Press.